CADAM 2014

SELECTED PAPERS

12th International Scientific Conference on Advanced Engineering, Computer Aided Design and Manufacturing – CADAM 2014

Boris Obsieger, Editor

Published in Croatia and USA, September 2014.

CADAM 2014
Selected papers

12th International Scientific Conference on Advanced Engineering,
Computer Aided Design and Manufacturing – CADAM 2014

Editor:
Prof. Boris Obsieger, D.Sc.,
University of Rijeka

Assistant editor:
Branimir Rončević, D.Sc.,
University of Rijeka

Reviewed by:
International Reviewing Committee
Prof. Iztok Potrč, D.Sc., president
University of Maribor

Typesetting:
Branimir Rončević, D.Sc.

Cover design:
Prof. Boris Obsieger, D.Sc.

Published by the editor

Print and distribution by Redak and On Demand Books
ISBN 978-953-7919-48-1 (soft cover, black and white)
Available at knjizara.hr and The Espresso Book Machine® channel.

Print and distribution by Redak and LightingSource
ISBN 978-953-7919-47-4 (soft cover, colour)
Available at knjizara.hr, from from Ingram, Baker&Taylor, Bertrams and other retail outlets.

Print and distribution by CreateSpace   *and CS eStore*
ISBN 978-1501039010 (soft cover, black and white)   at createspace.com/4980312
ISBN 978-1502366436 (soft cover, colour)   at createspace.com/4990688
Available from CreateSpace, Amazon and other retail outlets.

eBook edition for Adobe® Digital Editions supported devices
ISBN 978-953-7919-49-8 (colour)
Available from Ingram Digital and other retail outlets at all continents.

CIP: 130508038
CIP catalogue records are available from the University Library in Rijeka

Avaliable at **Australia, Brazil, Canada, Europe, Russia, USA** and many other countries at all continents.

International Scientific Conference on Advanced Engineering, Computer Aided Design and Manufacturing – CADAM 2014

September 16th – September 20th 2014, Vodice – Croatia

Organised by

University of Rijeka
Faculty of Engineering

University of Split, Faculty of Electrical
Engineering and Naval Architecture

University of West Bohemia Plzeň
Faculty of Mechanical Engineering

University of Osijek
Faculty of Electrical Engineering

Technical University of Sofia
Faculty of Management

University of Miskolc, Faculty of
Mechanical Engineering and Informatics

University of Zagreb
Faculty of Graphic Arts

University of Applied Sciences
FH JOANNEUM, Graz

International Institute
Info-Ruthenia, Moscow

Moscow State Industrial University

University of Maribor
Faculty of Mechanical Engineering

University of Mostar, Faculty of
Mechanical Engineering and Computing

Logistics

REVELIN
KONGRESNA AGENCIJA
Revelin d.o.o. Ičići
http://revelin.hr

INTERNATIONAL ORGANIZING COMMITTEE

Boris Obsieger, *President*
Vaclava Lašová, *Vice-president*

Ivan Dakov
Damir Jelaska
László Kamondi
Valery Koskin
Branka Lajić
Srete Nikolovski
Milenko Obad
Aneliya Petkova
Valery Poroshin
Iztok Potrč
Branimir Rončević
Domagoj Rubeša
Matjaž Šraml
Ctibor Štádler
Agnes Takács
Vojo Višekruna

Mladen Milinović, *Executive Secretary*

INTERNATIONAL SCIENTIFIC COMMITTEE

Iztok Potrč, *President*

Dmitry Bogomolov
Roman Čermák
Srečko Glodež
Damir Jelaska
László Kamondi
Vaclava Lašová
Tone Lerher
Valery Lyssenko
Gordana Marunić
Srete Nikolovski
Boris Obsieger
Valery Poroshin
Evgeny Pushkar
Anatoly Sheypak
Dubravka Siminiati
Matjaž Šraml
Vojo Višekruna

LIST OF PARTICIPANTS

Bertović, D., Rijeka, Croatia
Bogomolov, D., Moscow, Russia
Chicheryukin, V. N., Moscow, Russia
Chivileva, M., Moscow, Russia
Čermák, R., Plzeň, Czech Republic
Gregov, G., Rijeka, Croatia
Grishin, A. I., Moscow, Russia
Jelaska, D., Split, Croatia
Lašová, V., Plzeň, Czech Republic
Limberg, L., Plzeň, Czech Republic
Lovrin, N., Rijeka, Croatia
Lyssenko, V., Moscow, Russia
Ljubič Mrgole, A., Maribor, Slovenia
Messineva, N., Moscow, Russia
Nizhnik, V., Moscow, Russia
Novikov, P. V., Moscow, Russia

Obsieger, B., Rijeka, Croatia
Perkušić, M., Split, Croatia
Podrug, S., Split, Croatia
Poroshin, O., Moscow, Russia
Poroshin, V., Moscow, Russia
Potrč, I., Maribor, Slovenia
Rakitin, J., Moscow, Russia
Rončević, B., Rijeka, Croatia
Sarka, F., Miskolc, Hungary
Sever, D., Maribor, Slovenia
Sheypak, A. A., Moscow, Russia
Siminiati, D., Rijeka, Croatia
Štádler, C., Plzeň, Czech Republic
Takács, Á., Miskolc, Hungary
Vrcan, Ž., Rijeka, Croatia

CONTENTS

CADAM 2014

INVITED LECTURE

12th International Scientific Conference on Advanced Engineering, Computer Aided Design and Manufacturing – CADAM 2014

September 16th – September 20th 2014, Vodice – Croatia

Invited lecture by

Prof. Neven Lovrin, D. Sc.
University of Rijeka, Croatia
Faculty of Engineering

SOME ETHICAL ASPECTS OF ENGINEERING PROFESSION

Lovrin, N.

Abstract: Engineers bear the responsibility for the quality of life of future generations more than ever before. The decisions and actions of engineers have a profound impact on the world we live in, and society at large. Engineers have to be aware of the fact that by using available engineering technologies it is possible to provide abundance for all human beings, but also to destroy all life on Earth. Clear understanding of engineering ethics is needed like never before, because when things go wrong, there is always an ethical dimension.

Keywords: engineering profession, responsibility, engineering ethics, technology.

1 INTRODUCTION

The end of the twentieth and the beginning of the twenty-first century are marked by developments in science which is considered to be the basis of the greatest quantitative and qualitative changes in history. We are witnesses to the great benefits to mankind stemming from contemporary engineering development. The nuclear and space age that we live in, encourages the vigorous progress of science. Human technologies are developing very fast. Mechanization, automation and computerization of production processes have lessened the hazards to human physical integrity, but in spite of that, man's psychic and moral integrity in his working environment has been increasingly endangered. Modern technology has a deep impact on humankind and all life on Earth. Unfortunately we frequently are witnesses of more and worse or even tragic conse-quences of scientific and technological advances markedly caused by neglecting moral principles in people's activities.

In the modern era, engineering profession is no longer a pure technical discipline. The decisions and actions of engineers seriously affect the world we live in, and soci-ety at large. Therefore it is no longer possible to practice engineering without regard for the ethical context. Engineers have to be aware of their responsibility, dignity and ethics as they make choices during their professional practice and they should not think only about profit. Therefore, a clear understanding of engineering responsibility, dignity and ethics is needed like never before [1]. Modern engineers have to study and apply ethical codes, doctrines and principles in their professional engineering practice [2]. Examples of possible ethical dilemmas, regarding engineering profession, that may occur, are discussed on the following case.

2 THE SPACE SHUTTLE CHALLENGER DISASTER CASE

On January 28, 1986, seven astronauts were killed when the space shuttle they were

9

aboard of, the Challenger, exploded just over a minute (73 seconds) into the flight (Figure 1). The failure of the solid rocket booster O-rings to seat properly allowed hot combustion gases to leak from the side of the booster and burn through the external fuel tank. The failure of the O-ring was attributed to several factors, including a faulty design of the solid rocket boosters, insufficient low-temperature testing of the O-ring material and the joints that the O-ring sealed, and a lack of proper communication between different levels of NASA management.

Fig. 1. Launch (left) and explosion (right) of the Space Shuttle Challenger

The Space Shuttle Challenger disaster was a preventable tragedy and NASA tried to cover it up by calling it a mysterious accident. However, one man had the courage to bring the real true story to the eyes of the public and it is to Roger Boisjoly to whom we are thankful. Many lessons can be learned from this disaster to help prevent further disasters and to improve on engineering ethics.

Roger Boisjoly [3], the chief O-ring engineer, who worked for Morton Thiokol, the company which manufactured the booster rockets, knew the problems with the O-ring all too well (Figure 2). More than a year earlier he attempted to warn the panel of vice presidents of the inherent dangers involved in the launch. But despite his warnings and anxiety of some others Thiokol engineers and managers, Challenger was launched. The question was: Why, on the eve of the Challenger launch, did NASA managers decide that launching the mission in such low temperatures was an acceptable risk, despite the concerns of their engineers regarding problems with the O-ring? Possible answers could be: urgency due to the competition with Russians to be the first to observe Halley's Comet or some kind of political and/or economical pressure from the USA government or Congress.

Roger Boisjoly, the Thiokol engineer, testified before the Congress and sued the Morton Thiokol company under a federal whistleblowing statute, but he lost. Thiokol gave up on a ten million dollars incentive fee, but did not sign a document admitting to legal liability. After all the testimonials Biosjoly was taken off the project and subtly harassed by the Thiokol management. He left the company and underwent therapy for a post-traumatic stress disorder. For his honesty and integrity leading up to and directly following the shuttle disaster, Roger Boisjoly was awarded the Prize for Scientific Freedom and Responsibility by the American Association for the Advancement of

Science. Roger Boisjoly now lectures on changing workplace ethics issues, a subject on which he has also spoken at MIT. Boisjoly said that this fatal accident could have been prevented. "The answer to preventing future catastrophes," he said, "lies in engineering ethics". The Challenger case is an excellent example showing that ethical issues involve confronting engineering responsibility against management decision-making.

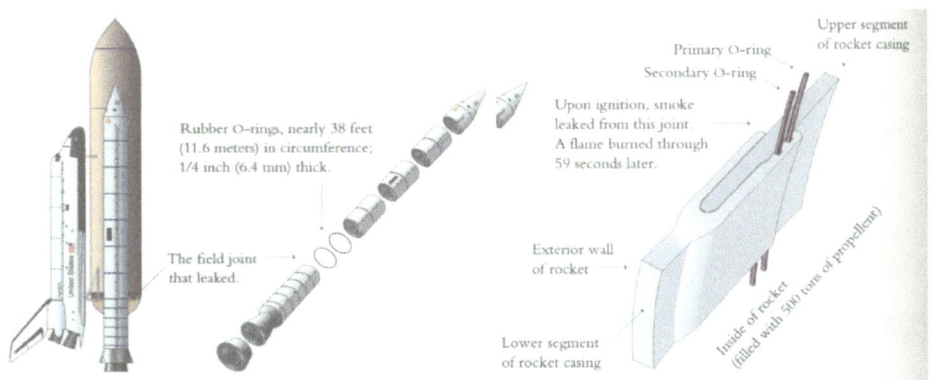

Fig. 2. Design of the field joint with rubber O-rings

3 ENGINEERING ETHICS IN MODERN SOCIETY

Engineers often face the dilemma of loyalty to their company and employer versus their responsibility to the society and environment. The dilemma is whether a product presents an enormous hazard, whether engineers have a duty through their individual consciousness to make the problem public or whether they have to protect their company.

Engineers have obligations to future generations that could be harmed by irresponsible engineering activities, because it may take decades and generations for products and facilities to have adverse effects. Engineers who draw attention to problems against the wishes of their superiors are known as whistleblowers. They can expect that being a whistleblower will have serious consequences for their job and perhaps for the rest of their career – even when the problem they are drawing attention to is real and important. But if an engineer's superior, manager or colleague does not act to undo, curb or mitigate dangers of serious personal or social damage, it is necessary to offer honest criticism or even "blow the whistle", so as to reduce the risk, to acknowledge and correct errors, violations or negative consequences as much as possible [4].

Engineers should not act using immoral and unethical rules and laws. They should not be bribed and corruptible. Engineers should always keep in mind the moral responsibility and obligations toward society as a whole. Their professional ethical standards have to transcend commonly accepted morality [5]. Ancient Chinese people summarized all these highest virtues in three Chinese words: Zhen (truth, truthfulness), Shan (kindness, benevolence, compassion) and Ren (endurance, forbearance, tolerance) [6].

Engineers of high ethical standards should not be involved in design of weapons for mass destruction i.e. chemical, nuclear and biological weapons [7]. They also should not be involved in so called researches, which are in reality paid by various companies to advertise their products.

Modern engineers also have to develop spiritual intelligence [8]. Spiritual intelligence can be described symbolically as the backbone of human consciousness, responsible for character building and meaning making. Developing spiritual intelligence is more of an experiential rather than a theoretical process. The language of spiritual intelligence is the language of the heart. Growing in spiritual intelligence, engineers grow in their action logic from the perception of "What I can get ..." to "What I can contribute ... ". The practice of self-reflection and contemplation enhances development of spiritual intelligence, and a depth of compassion and benevolence to all life on Earth develops as well. Thus, modern engineer will develop the ability to act with wisdom and compassion, while maintaining inner and outer peace (equanimity), regardless of the circumstances.

To accomplish all this goals and to be a engineer of a high quality, modern engineers have to study, not only engineering, but also ethics and philosophy in order to understand relationships between man, nature and the universe and thus to become a humanist who respects, protects and welcomes all life on Earth.

4 CONCLUSION

Engineering ethics is a crucial point and essential for our survival. It is not an option or a luxury. Engineers have to be aware of ethics as they make choices during their professional practice. Engineers must perform under a standard of professional behavior that requires adherence to the highest principles of ethical conduct including honesty, impartiality, fairness, and equity, and do so in the absence of bribe and corruption. Modern engineers have to study and apply ethical codes, doctrines and principles in their professional engineering practice. They have to recognize the importance of sociological and cultural context of the engineering profession. They should also contribute to environmental protection and to sustaining the balance in nature.

References:
[1]   Verein Deutscher Ingenieure. *Fundamentals of Engineering Ethics,* Düsseldorf, 2002.
[2]   Mance, K. (2007). *Engineering Ethics Teaching*, Engineering Review, Vol. 27 No. 2, Faculty of Engineering, University of Rijeka, Rijeka, Croatia.
[3]   Boisjoly, R. *The Challenger Disaster*, Online Ethics Center for Engineering, National Academy of Engineering, Washington, D.C., USA, 2006.
[4]   Baura Gail, D. *Engineering Ethics: An Industrial Perspective*, Elsevier Academic Press, USA, 2006
[5]   *Proceedings of the Round table discussion, Ethics in Application and Development of the Engineering Sciences*, Croatian Academy of Engineering, Zagreb, Croatia, 2005.
[6]   Hongzhi, L. *Zhuan Falun*, Fair Winds Press, Gloucester, Massachusetts, USA, 2001.
[7]   Bowen, R. W. *Engineering Ethics*, Springer-Verlag London Limited, 2009.
[8]   Zohar, D. & Marshall, I. *SQ: Connecting with Our Spiritual Intelligence*, Bloomsbury Publishing, New York and London, 2000.

CADAM 2014

SELECTED CONFERENCE PAPERS

12th International Scientific Conference on Advanced Engineering, Computer Aided Design and Manufacturing – CADAM 2014

September 16th – September 20th 2014, Vodice – Croatia

Authors

Bertović, D.; Bogomolov, D.; Čermák,R.; Chicheryukin, V. N.;
Chivileva, M.; Gregov, G.; Grishin, A. I.; Jelaska, D.; Lašová,V.
Limberg, L.; Lovrin, N.; Messineva, N.; Mrgole, A.; Nizhnik, V.
Novikov, P. V.; Lyssenko,V.; Perkušić, M.; Podrug, S.;
Poroshin, V.; Rakitin, J.; Sever, D.; Sheypak, A. A.; Siminiati, D.
Takács, Á.; Vrcan, Ž.

SIMULATION MODEL OF A SERIAL HYDRAULIC HYBRID DRIVE TRAIN FOR A FORKLIFT TRUCK

Bertović, D.; Gregov, G. & Siminiati, D.

Abstract: Global problems such as environmental pollution and limited reserves of fossil fuels, demand the increase energy efficiency on a conventional vehicle. One idea for solving this problem is development of hybrid technologies. Hybrid vehicle is a vehicle that can use two or more systems to create, store and reuse the energy. In this paper has been carried out calculation of hydraulic transmission for forklift truck. After that has been developed simulation model of serial hydraulic hybrid system for the same forklift truck using the ITISim software.

Keywords: *hybrid hydraulic system, simulation model, forklift truck.*

1 INTRODUCTION

The definition of hybrid vehicles is: a vehicle that can use two or more systems to create, store and reuse the stored energy [1]. The idea of a hybrid vehicle has been developed from the need to increase efficiency on a conventional vehicle, thus is necessary to incorporate additional systems on it. This creates a vehicle with primary and secondary system. The primary system is used to power the vehicle, a secondary system is used to produce, store and reuse energy which would be irrevocably rejected in the environment from the primary system. Losses in conventional vehicles such as losses in the engine, power transmission, as well as losses due to air resistance and rolling are inevitable. However, the energy that is lost during braking can be stored and reused, thus increasing the overall efficiency of the vehicle. This principle of utilized braking energy is called regenerative braking. Depending on design of the secondary system (operating principle), hybrids are divided on electrical, hydraulic and mechanical hybrid.

2 HYDRAULIC HYBRID SYSTEMS

The development of hydraulic hybrid has begun in the 1980s of the last century. Volvo Flygmotor has developed parallel hybrid, which is installed on buses in Stockholm. The reduction of the fuel consumption was between 16% and 25% [2]. The main components of the hydraulic hybrid systems are hydraulic pump, hydraulic motor and hydraulic accumulator which are equivalent of the generator, electric motor and electric battery in electric hybrids. The main advantage of the hydraulic hybrids compared to electric hybrids, is a greater amount of stored energy density which is up to 5 times higher [3].

The working principle of the hydraulic hybrid is as follows: when the vehicle starts to break, hydraulic pump connected to the drive wheels establishes the flow of the working fluid from the low pressure accumulator to high pressure accumulator. The fluid must overcome the pressure of the gas, which is currently acting in the high-pressure accumulator. Compressed gas creates resistance to the flow, which creates the effect of braking on the wheels and slows the vehicle. Thus, the kinetic energy of the vehicle is converted into energy of compressed gas in the accumulator. When the vehicle begins to accelerate, the working fluid through the pump is diverted from the accumulator to low pressure reservoir. Now the pump is acting as a hydraulic motor. When the vehicle consumes all stored energy, the drive from the secondary system switches to the primary.

The basic division of the hydraulic hybrid is on the serial and parallel (Fig. 1). Typical serial hybrids are mounted on vehicles that already have a hydrostatic transmission, such as delivery vehicle which driving regime is made up of a lot of stopping and starting. The lack of a serial hybrid is the inability to achieve high-speeds, making them ideal for use on delivery vehicles, but not for vehicles in intercity traffic. For such vehicle is more convenient to use a parallel hybrid system, because it allows the storage of energy during braking, but when they need to achieve higher speeds they use the primary system.

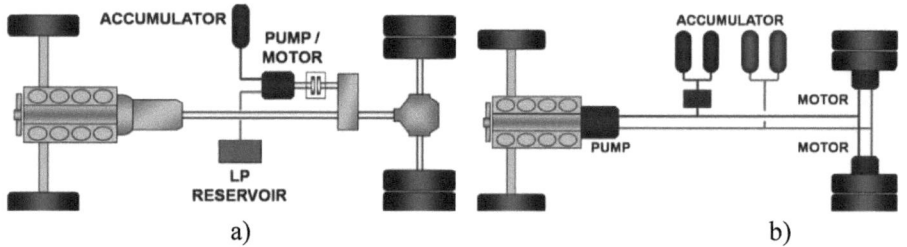

a)   b)

Fig. 1. a) Parallel hydraulic hybrid vehicle, b) Serial hydraulic hybrid vehicle

3 CALCULATION OF A HYDRAULIC SYSTEM FOR A FORKLIFT TRUCK

Calculation of a serial hydraulic hybrid system is carried out for a vehicle that uses hydrostatic transmission. Forklift is chosen primarily because its driving regime has a lot of stopping and starting and for simplify the calculation of the required parts. So the vehicle already has a hydrostatic transmission and some usable specifications are known. As a model for the calculation, the forklift KOMATSU FH 50 was chosen [4].

Hydrostatic system of a forklift truck consists of two hydraulic motors, mounted directly on the drive wheels via planetary gearbox, while the pump is connected to the IC engine. The first step in the calculation is to estimate all the resistances that occur when the forklift is moving. Since the maximum speed of the forklift is 23,5 km/h the air resistance is ignored. Therefore the only remaining resistance is the rolling friction force. Consider an average friction coefficient for asphalt $f = 0,019$, rolling friction force $F_{FR} = 1918$ N has been obtained. The torque that occurs on wheels is $T_{FR} = 525$ Nm so when you divide it on two wheels is $T_W = 262,5$ Nm. Transmission ratio of the planetary gearbox has been chosen $i_{PP} = 5$, so the torque that each hydraulic motor must achieve is $T_{HM} = 52,5$ Nm. Depending on the minimum (5 km/h) and maximum speed (23,5 km/h) of the truck, resulting rotational wheel speeds are $n_{min} = 48,5$ min^{-1} and $n_{max} = 227,9$ min^{-1}. Assuming the pressure drop $\Delta p = 250$ bar and the hydro-mechanical efficiency $\eta_{HM} = 0,95$ gives displacement of hydraulic motor of $V_{HM} = 14$ cm^3. According to the obtained value for displacement, axial piston fixed hydraulic motor Bosch Rexroth A4FM [5] has been chosen. Hydraulic motor with larger displacement was chosen because of the needs of a hydraulic accumulator charging during regenerative braking. The maximum flow of the selected hydraulic motor at the speed of $n_{max} = 227,9$ min^{-1} is $Q_{HM,ef} = 26,2$ l/min which means that the flow of the pump must be the minimum $Q_{HP,ef} = 52,4$ l/min.

Based on the required pump flow and the optimum rotational speed of the engine $n_M = 1400$ min^{-1} and hydro-mehanical efficiency of the pump $\eta_{HP} = 0{,}97$, pump displacement should be $V_{HP} = 36$ cm^3. According to the obtained value for displacement, the axial piston variable pump Bosch Rexroth A4VSO [5], has been chosen.

4 SIMULATION MODEL OF A SERIAL HYDRAULIC HYBRID SYSTEM

To develop a simulation model of the serial hydraulic hybrid the software ITISim was used. The first model represents only the energy storage mode. The model consists of two hydraulic motors and hydraulic accumulator. The model also included block which represents a speed decreasing. Simulation of energy storage is defined as a uniform linear deceleration of vehicle from 20 km/h to 16 km/h during 20 s. This mode is the deceleration of the vehicle while charging the high pressure accumulator. For instantaneously breaking of the vehicle the mechanical brake is used.

After several simulations (changing accumulator volume and maximum pressure) satisfactory results were obtained. The final volume of the accumulator is 20 l and 250 bar charging pressure. The slowdown in this simulation is assumed to be at an average speed of the truck, and realized braking time of 14 s, fulfil the above criteria. According to this result (Fig. 2a) the accumulator was charged with 10 l of oil and the obtained charging pressure of 250 bar (Fig. 2b). Based on the results of optimization the Bosch Rexroth Bladder Type accumulator [5] was selected.

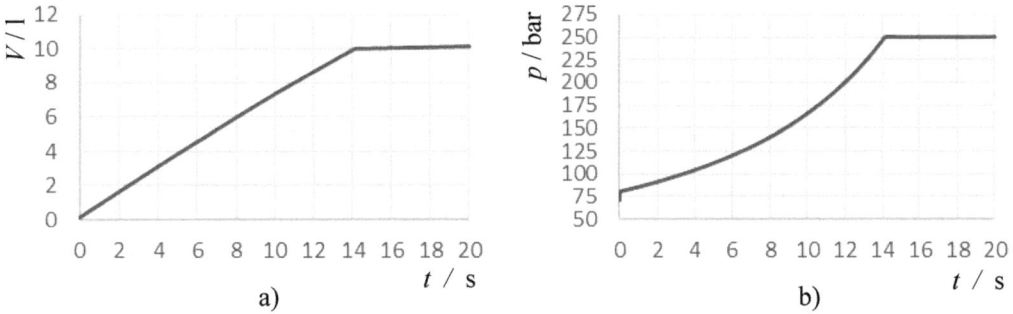

Fig. 2. Simulation results: a) accumulator charging volume, b) achieved pressure in accumulator

The next step was to make a simulation model that includes braking, starting and driving mode. In the previous simulation model has been added another reservoir of low pressure. Unfortunately in ITISim software there is no way to simulate braking and acceleration with the same block. So the first simulation model has been extended with two additional hydraulic motors, which will be used to simulate the start of the vehicle. The developed model consists of valve which is closed after high-pressure accumulator is charged, so that the oil cannot flow back to the first pair of hydraulic motors after the beginning of start mode. Thereafter, another valve opens and diverts oil to the low-pressure accumulator through the second pair of hydraulic motors which simulates the starting of the vehicle. After the end of the starting mode it is necessary to switch the secondary system to the primary to power the vehicle, so the hydraulic pump is turned on.

Starting vehicle simulation results are presented in Fig. 3a, which shows the increase in speed when starting the vehicle from standby mode. At that point there is a sudden increase of rotation speed of the hydraulic motors, due to the starting of the vehicle. The results show that the rotational speed of the forklift wheel rises to 64 min^{-1}, after which the speed decreases. The conclusion is that the value of the speed depends on the high-pressure accumulator volume and pressure.

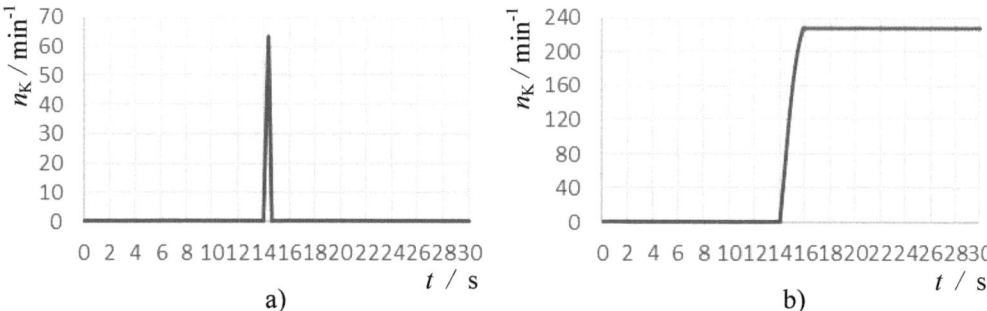

Fig. 3. Wheel rotation speed: a) during the forklift start, b) during the vehicle driving

After the initial start-up the vehicle, the hydraulic pump is activated thus the hydro-static transmission continues to run the vehicle. In order to avoid the shock caused due to differences in the speed of rotation the pump is turned on at the 14,5 s. Results of the rotation speed (Fig. 3b) shows that between 14 s and 14,5 s the vehicle accelerates using the stored energy and then continues to accelerate to full speed using a hydro-static transmission. It can be concluded that between starting and driving mode has been achieved continuous speed changing. During driving mode has been obtained the maximum rotation speed of the forklift wheel of 227 min^{-1}, as well.

5 CONCLUSION

The main objective of this study was to develop a simulation model of serial hydraulic hybrid system for the forklift truck, using the software ITISim. Simulations of the model have been performed for braking, starting and driving mode. Results of simulation for breaking mode showed, that during the deceleration of vehicle speed, the high pressure accumulator charged with adequate volume of working fluid. This volume of oil is enough for starting the vehicle in start mode. Between starting and driving mode has been achieved continuous speed changing. During the driving mode has been obtained the max. rotation speed of the forklift wheel as well. Finally, it can be concluded that the results of the analysis indicate to the real behaviour of a developed simulation model.

References:
[1] http://www.thefreedictionary.com/hybrid, 16.07.2014.
[2] Ibrahim, M. S. A. *Investigation of Hydraulic Transmissions for Passenger Cars*, doctoral thesis, Aachen, Germany, 2011.
[3] Rydberg, K. E. *Energy Efficient Hydraulic Hybrid Drives*, The 11th Scandinavian International Conference on Fluid Power, SICFP'09, Linköping, Sweden, 2009.
[4] Yamamoto, H.; Harad, Y. & Hiraiwa, H. *Introduction of Hydrostatic Transmission Forklift Model FH40-1/FH45-1/FH50-1*, Komatsku tehnical report, Vol. 58, No. 165, 2012.
[5] http://www.boschrexroth.com/en/xc/23.07.2014.

MODELLING OF FLOW AND HEAT FLUX IN THIN 2D CHANNEL WITH ROUGH MOVING WALLS

Bogomolov, D.; Poroshin, V. & Nizhnik, V.

Abstract: Paper describes the mathematical model of flow and heat flux in thin channel with rough moving walls in the 2D approach. The effect of the real measured roughness profiles upon Nusselt number in comparison with smooth wall channel is shown. Both static and dynamic flow components are investigated.

Keywords: *thin channel, surface roughness, fluid flow, heat flux, moving wall.*

1 INTRODUCTION

Study of the fluid flow and heat flux in thin channels of technological devices is one of the main problems of modern applied mechanics. Practical aspects of the problem are referred to the heat flux intensification in compact heat exchangers, heat flux study in moving parts of pumps and engines and so on. Behaviour of the flow and heat flux in the thin channel differs from the same processes in large size channels. Due to the small gap size, surface roughness becomes one of the main factors, having great impact on the resulted leakage and heat-exchange rate in the channel. One of the common ways to stimulate heat-exchange in large channels and pipes (especially for turbulent flow) is to generate is to produce some regular surface texture on its' walls. In the small channel surface texture influence should be even more clear.

In authors' earlier work [1] the mathematical model of heat flux in the thin channels with rough immovable walls were proposed. Present work is an attempt for extending our researches over the heat flux in channels with rough moving walls. Such model allows to analyze the roughness effect upon the static and dynamic components of flow and heat flux in the thin channel.

2 CALCULATION MODEL

The geometry model of the thin channel with moving walls formed by two rough surfaces h_1, h_2 is presented in Fig. 1. The gap between two rough surfaces (H) is taken as the distance between their mean lines. Upper and lower walls has coordinates $H_1(x) = H + h_1(x)$ and $H_2(x) = H + h_2(x)$. The current gap in the channel is calculated as $h_T(x) = H + h_1(x) + h_2(x)$. The upper wall slides in parallel direction to the lower with constant U speed.

Measured profiles of the real industrial surfaces or simulated roughness profiles with desirable features can be used as a surface geometry. Both profiles are always specified on regular grid with Δx step. Calculation step in x direction is considered as some fraction of the profile grid step $\Delta x = k \delta x$ (where k is an integer value). Calculation step in y direction δy. is also considered regular comparable to δx.

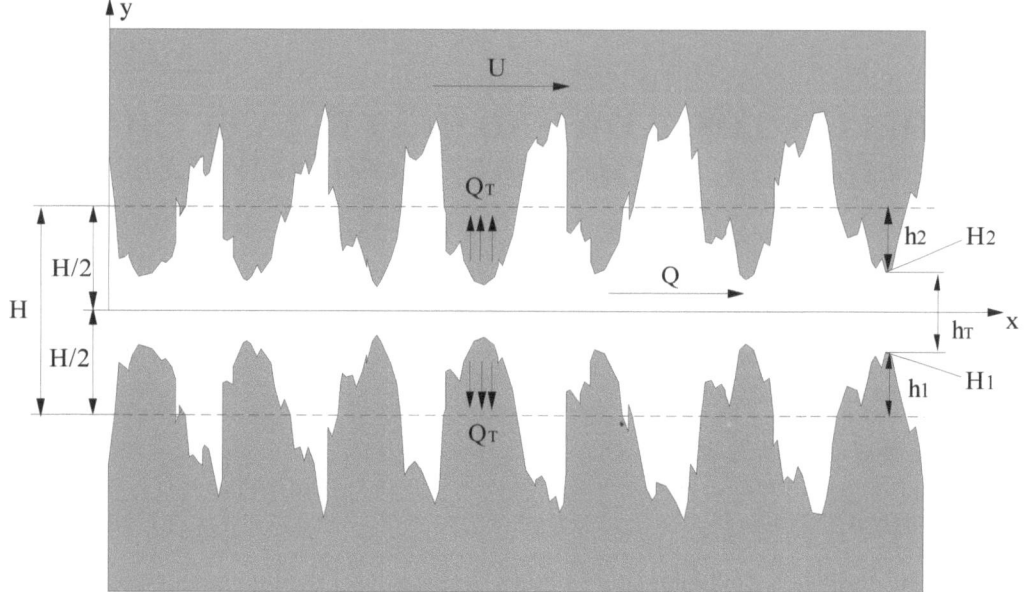

Fig. 1. Geometry model of the thin channel with rough moving walls

To model the flow in the described thin channel the lubrication flow model was proposed. It is based on the common Reynolds equation, which can be solved analytically for the 2D case [2].

$$p(x) = p_{in} + 6\mu U \int_0^x \frac{d\xi}{h_T^2(\xi)} + \frac{p_{out} - p_{in} - 6\mu U \int_0^L \frac{d\xi}{h_T^2(\xi)}}{\int_0^L \frac{d\xi}{h_T^3(\xi)}} \int_0^x \frac{d\xi}{h_T^3(\xi)}, \tag{1}$$

where p is the flow pressure, p_{in} is the pressure at the channel inlet, p_{out} is the pressure at the channel outlet (boundary conditions), μ is the dynamic viscosity.

To model the heat flux the thermal balance equation is commonly used. Supposing the process as stationary and assuming that the flow is in general ordered in the x direction:

$$\rho C_p v_x \frac{dT}{dx} = \lambda \frac{d^2 T}{dy^2}, \tag{2}$$

where T is the local flow temperature, ρ is the media density, C_p is the media thermal capacity, λ is the media thermal conductivity, v is the local flow velocity.

Difference problem (2) was solved with known temperature T_w at the channel walls and known temperature of the flow T_{in} at the channel inlet. Under such boundary conditions the problem was solved by using the implicit finite-difference method. The sweep method (tridiagonal matrix algorithm) was used for the solution.

The operational capacity of the thin channel with rough walls was assessed by means of Nusselt number Nu. The static (no wall movement) and dynamic (no pressure gradient) boundary conditions were used for analyse the influence of both static and dynamic components of the flow.

3 ANALYSIS

To verify the developed calculation model and computation software the channel with smooth walls was examined, which has already been studied by many authors [3]. The results of the numerical experiments are agreed to the expected process behaviour. The quasy-parabolic distribution of the temperature was achieved for the inner sections of the channel. There is some zone near the channel inlet where the process is developing (the cross section of the temperature field has a shape of flattened parabola).

Other numerical experiments were carried out for the rough wall channels. The surfaces after polishing, grinding, milling and turning were used. The cross sections of the flow temperature fields are also has a quasi parabolic shape as for the smooth channel. The heat flux is more intensive for the greater roughness heights.

The resulting Nusselt number evolution for flows in channels with different surface roughness for static and dynamic boundary conditions are shown in Fig. 1, 2.

Nusselt number grows with growing average gap, because Reynolds number of the flow also grows. It is consistent to the known theoretical assumptions and experimental data for macro-size channels.

For both static and dynamic flow components the Nusselt numbers for the rough surfaces becomes more distinct from the smooth channel ones with growing of the average gap. But the difference in Nusselt numbers is much lesser for the more thin channels.

For the static case the mean flow temperature is located approximately in the middle of the channel vertical section. The central flow layers are mostly affect the resulting Nusselt number.

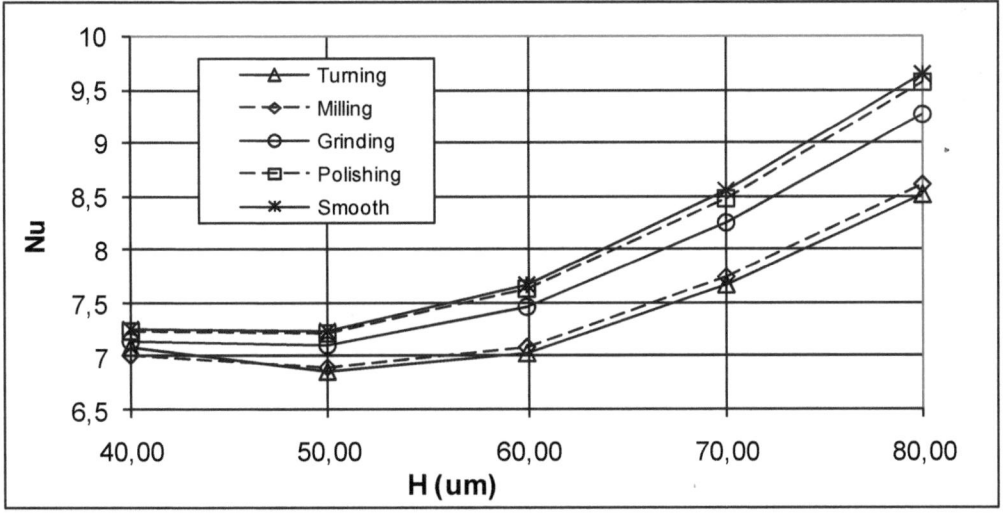

Fig. 1. Evolution of the Nusselt number for static flow component

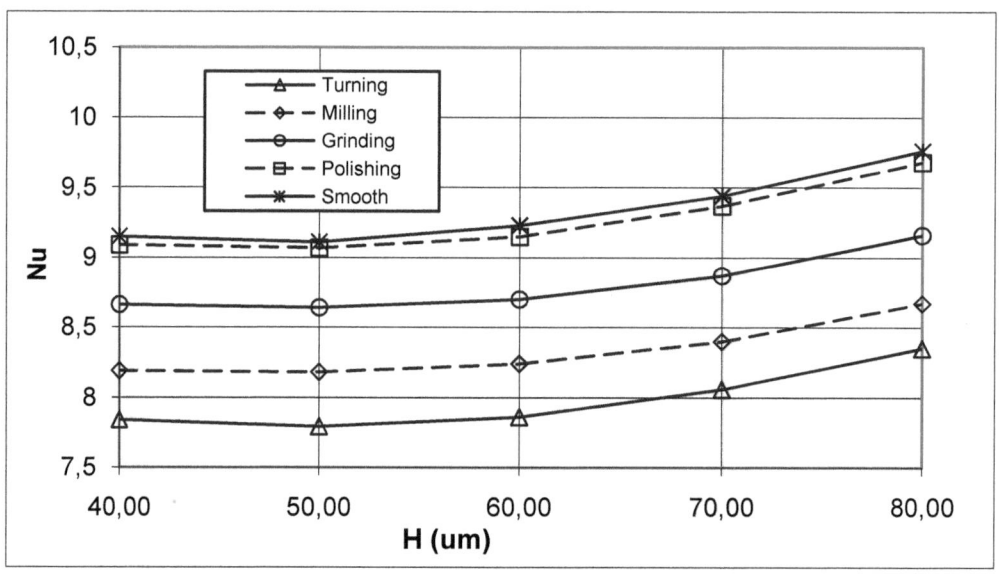

Fig. 2. Evolution of the Nusselt number for dynamic flow component

For the dynamic case the mean flow temperature is biased in the direction of the moving wall. The layers adjacent to the moving wall are much more involved into the heat flux. The Nu curves for the dynamic case are quite parallel to each other. That is an unsuspected fact that requires further detailed investigation.

4 CONCLUSION

The proposed mathematical model allows to forecast flow and heat flux in 2D channels with rough moving walls. It can be effective for the optimal selection of the processing methods of surfaces at the projecting stage. Numerical experiments show that the surface roughness have a great influence upon the heat flux. For both static and dynamic flow components the Nusselt numbers for the rough surfaces becomes more distinct from the smooth channel ones with growing of the average gap.

The research was performed with the financial support of the Ministry of Education and Science of the Russian Federation for higher education institutions within the state job service.

References:
[1]   Bogomolov, D.; Poroshyn, V.; Lysenko, V. & Onanko A. (2011). *Modeling of flow and heat transfer in thin 2D channels with rough walls*, Advanced engineering, Vol. 5 No. 2, pp. 147-154.
[2]   Sheypak, A.; Poroshin, V.; Syromiatnikova, A. & Bogomolov, D. (2008). *Roughness influence upon the hermeticity of plunged pair using equivalent gap model*, Advanced Engineering, Vol. 2 No. 2, pp. 283-290.
[3]   Mills, A. F. *Heat Transfer, 2nd Ed.* Prentice Hall, Upper Saddle River, New Jersey, 1999.

ELECTROMAGNETIC DAMPER AND ITS APPLICATIONS

Čermák, R.

Abstract: The contribution deals with the use of passive, active and semi-active dampers, esp. electromagnetic dampers, in automotive applications. Principles of dampers and their applications are discussed. More focus is given to electromagnetic damper and its design. Simulation possibilities in various commercially available SW packages are mentioned. Several simulations and their results are discussed. Also a prototype of the damper is mentioned.

Keywords: suspension, shock absorber, eddy currents, electromagnetic damper.

1 INTRODUCTION

Automotive suspension is one of the key components in nowadays cars. Proper suspension design can significantly improve vehicle behaviour and safety.

The analysis of passive automotive suspension in literature (e.g. [1], [2], [3], [4]) shows, that there are significant trade-offs in performance between the ride quality, rattle space and tire deflection transfer functions. Improvements in one of them usually deteriorate the other two transfer functions. Therefore active and semi-active suspensions were introduced.

The fundamental difference between active and semi-active control strategy is in the form of input energy. While active suspension uses force actuator to control motion of the body (i.e. it needs enough energy to move the body with the desired velocity and acceleration), semi-active suspension system utilizes a variable damper or other dissipation component (see Fig. 1). Compared to active suspension, semi-active one consumes significantly less power.

2 SUSPENSION MODELING

Short introduction to the suspension modeling is given in this section.

2.1 Quarter-car, half-car and full-car models

Simplified models are very often used for the modeling of vehicle suspension. Typical example is using of a quarter-car model for description of vertical motion of vehicle. The model is very simple and gives relevant results. For more complex study (including roll and pitch motions, etc.) a half-car or a full-car model must be used.

Detailed mathematical description of the models can be found in the literature (e.g. [1], [2], [3], [4]). For complex manoeuvre simulation an integration of the suspension model into full car model is required. The complete full car model usually includes submodels of powertrain, driveline, tires, brakes, steering, etc. Development of such a complex model takes lots of effort, and it is easier to use one of the commercially available and proven solutions described in the following section.

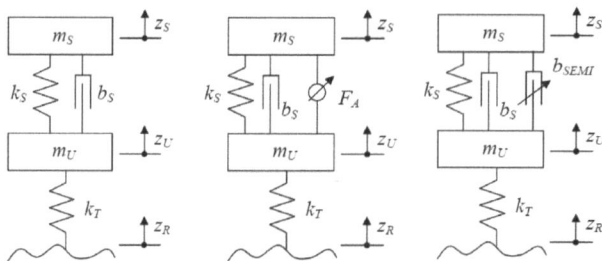

Fig. 1. A quarter-car model – from left to right: passive suspension, active suspension, semi-active suspension

2.2 Modeling of a full vehicle in ADAMS/Car

For overall understanding of vehicle behavior, it is useful to model the vehicle in one of available commercial software packages, which provide the user proven and quite complex models of vehicle components. Typical example of such a software package is ADAMS and its automotive modules called ADAMS/Car. ADAMS also offers possibility of co-simulation with software like MATLAB/Simulink and with EASY5.

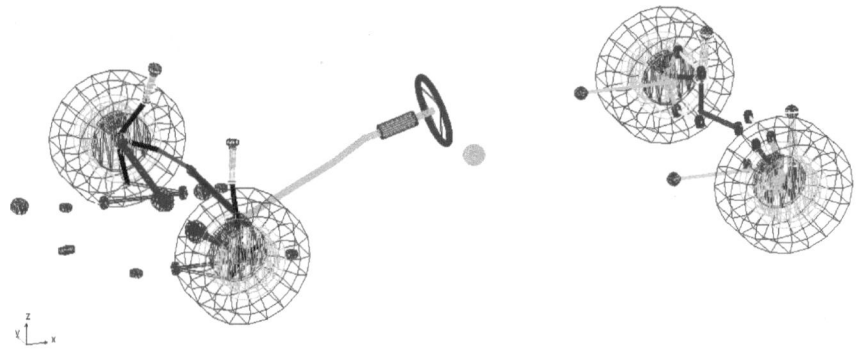

Fig. 2. Full vehicle model in ADAMS/Car package

ADAMS/Car provides the user with an easy way how the overall car model can be build. It uses so called template based approach, i.e. each component is described in a template, which is used to generate a subsystem as a part of the full vehicle assembly. Shock absorber properties are defined in a property file, an ASCII file with the force-velocity curve definition.

Among the above mentioned the SW provide user with a number of pre-defined manoeuvres and a tool – event builder – for creating user-defined manoeuvres. The SW contains a tool Road Builder for creating a test track with various obstacles (compliant road is also available). Many various functionalities such as Smart Driver, DoE and optimisation options are available in the SW.

There is an example of a road with several obstacles and a simulation results for a demo vehicle running across.

Complex simulation option of the full vehicle exposed to different road conditions, driver behaviour, etc., gives the users an excellent opportunity to investigate complex tasks in the virtual reality.

2.3 Electromagnetic shock absorbers

Although using of electromagnetic shock absorber is quite rare in the commercial applications, it is very often discussed in the scientific papers. There are several principles described in the literature.

One of the principles discussed is a passive electromagnetic damper based on eddy currents (see Fig. 3). A rod (mover) equipped with a set of neodymium permanent magnet and ferromagnetic poles is moving through a tube made out of a massive copper. The movement of the rod generates eddy currents in the copper tube and a force acting against the motion, which causes the required power dissipation. The force can be calculated either analytically or numerically. The principle of such a damper is shown in Fig. 3.

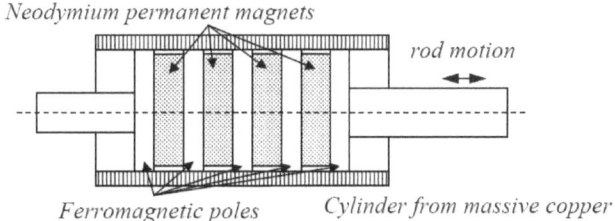

Fig. 3. Eddy currents based passive damper

A prototype of such a damper was designed (based on [7]) and manufactured at the author's workplace. The necessary calculations of the magnetic field were done in Comsol Multiphysics software.

Another principle of the electromagnetic damper is an arrangement similar to linear drive, so called tubular drive (see Fig. 3) – passive copper can be replaced by a stator winding or some mechanism can be used to convert linear to rotary motion. There are several principles mentioned in the literature using rotary actuators and planetary gear or various techniques to change linear motion of the rod to the rotary motion of the actuator. The actuator (either rotary or linear) is in principle an electric drive, either rotary or linear) containing a coil or several coils, which can be fed by constant or variable electric currents. Such an actuator can act as an active or semi-active damper and it also offers ability of energy recuperation. Energy recuperation option can be extremely interesting topic in nowadays electrical vehicles, because of large amount of energy dissipated in the damper during normal operation.

Combination of both principles mentioned above leads to a hybrid damper (see also [7]). Inspired by the text [7] a similar damper was designed at the author's workplace, and will be manufactured and tested in the nearest future.

3 TEST RIG DESCRIPTION

Mathematical models of the suspension require detail information about the shocks and their behaviour. Understanding of the shock absorber behaviour requires not only virtual testing (modeling and simulation), but also performing several physical tests. Therefore we have built a test device able to measure the basic characteristics of the shock absorbers, change properties of the semi-active shocks, understand the behaviour of the dampers and identify the parameters further used for simulation models (either simple quarter car model or the full vehicle model).

A simple structure (see Fig. 4) with four columns for fixing the damper was designed. The rod is driven by linear electric motor. Force sensor is located at the end of the rod. Displacement sensor is included in the drive. The damper characteristics are changed by means of current either in the solenoid valve (for DCC damper) or in the coils (for M-R dampers). Simple amplifier with IGBT transistor driven by analogous voltage from 0 to 10V is used to change the damping characteristics. The whole test rig is controlled from MATLAB via card MF624 and Real Time Target.

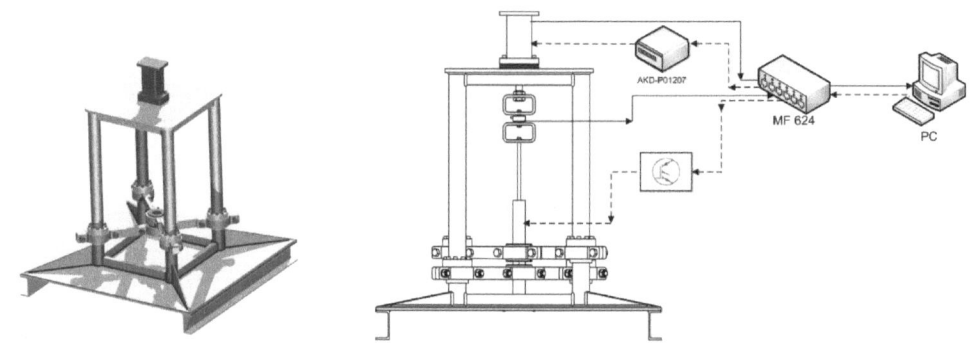

Fig. 4. Test rig for automotive shock absorbers [6]

4 CONCLUSION

The paper deals with application of electromagnetic dampers in automotive industry. Principle of an eddy current based damper is described. Test rig for shocks measurement and their parameters identification is briefly described. Modeling and simulation possibilities in commercially available software are briefly mentioned.

Acknowledgement:
This research is supported by internal grant of the University of West Bohemia in Plzen.

References:
[1] Guglielmino, E.; Sireteanu, T.; Stammers, C. W.; Ghita, G. & Giuclea, M. (2008). *Semi-active Suspension Control,* ISBN 978-1-84800-230-2, Springer, London.
[2] Savaresi, S. M.; Poussot-Vassal, C.; Spelta, C.; Sename, O. & Dugard, L. (2010). *Semi-active Suspension Control Design for Vehicles*, ISBN 978-0-08-096678-6, Elsevier Ltd.
[3] Rajamani, R. (2006). *Vehicle Dynamics and Control*, ISBN 0-387-26396-9, Springer, New York.
[4] Fijalkowski, B. T. (2011). *Automotive Mechatronics: Operational and Practical Issues, Vol.II*, ISBN 978-94-007-1182-2, Springer, New York.
[5] Dixon, J. C. (2007). *The Shock Absorber Handbook*, Second Edition, ISBN 978-0-7680-1843-1, John Wiley & Sons, Chichester, England.
[6] Slípka, F. (2012). *Active systems in vehicle chassis*. Diploma thesis, UWB Pilsen, Czech Republic *(in Czech)*.
[7] Ebrahimi, B. (2009). *Development of Hybrid Electromagnetic Dampers for Vehicle Suspension Systems*. Doctoral dissertation, University of Waterloo, Ontario, Canada.

PROPOSITION OF TRANSMISSIONS HAVING AN INDEPENDENTLY CONTROLLABLE OUTPUT SPEED

Jelaska, D.; Perkušić, M. & Podrug, S.

1 INTRODUCTION

A large portion of energy losses occur in power transmission and significant efforts are being made to improve the efficiency of transmission systems and reduce the production costs while keeping the optimal efficiency of the prime mover.

In vehicles, this aim is frequently tackled by continuously variable transmissions (CVTs) significantly improved by power-split CVTs [1,2] consisting of an ordinary CVT and a planetary gear train (PGT), and in wind turbine by hybrid transmissions [3-5]. The main problem of these transmissions is the high cost of the corresponding control system and power efficiency of the transmissions. These problems can be smoothly solved by transmissions with controllable output speed, independent of the input speed, so called independently controllable transmission (ICT). It consists of two PGTs connected with two gear drives: input one that splits the input power to PGTs and the output one which receives the parts of the PGT powers, sums them and sends to the output, termed the "free transmission shaft". One of the rest of PGTs powers flows to the control motor shaft and the other to the real output shaft, while the speeds of their shafts are linearly dependent. However, its "free shaft" runs at a variable speed synchronous with the input speed. It stays as a great shortcoming of that type of transmission.

2 PROPOSITION OF THE NOVEL ICT TRANSMISSION

A novel ICT is proposed herein based on transmission suggested by Hwang et al. [6]. To solve the problem of the free shaft power, it is required to direct the power flow of the free shaft at variable speed back into the input gear drive D, Fig 1. It could be done by simply connecting the gear drives E and D, i.e. by connecting the shafts S_6 and S_{in}, but in such a case one degree of freedom is lost and also the ability of controlling the output shaft speed. Therefore, one more PGT has to be added, PGT C, which then connects the gear drives E and D. Such transmission is found not to be feasible as well, because $\alpha = \beta = \gamma$, which leads to coupling points of all PGTs. To avoid this, the centre distances between middle shafts of gear drives should not be the same. Therefore, two more gear drives are added: the gear drive G between PGT C and the gear drive D, and gear drive H between PGT C and the gear drive F. The ICT is obtained now in which the power entering the input gear drive D is split into the power entering PGT A and that entering PGT B. One portion of these powers enters the output gear drive F and the other, over the gear drive F enters the PGT C, where that power is also split in two portions: one enters the output gear drive F over the gear drive H, whereas the other comes back to the input gear drive D over the gear drive G. In such a way, the power of gear drive G circulates in two closed circles, G–D–A–E–C–G and G–D–B–E–C–G, and never reaches an output shaft. The sum of the powers entering PGTs A and B is greater than the input power.

All shafts of gear drives D, E and G rotate at variable speed, while all shafts of gear drives F and H rotate at the speed linearly dependent on the control generator speed. Two output shafts of the gear drive F, which are at the same time output shafts of the entire transmission, are the control generator shaft, S_{cg}, and real output shaft, S_{out}.

The feasibility conditions for such type of ICTs are investigated. Twenty-one designs of PGTs A, B and C are found to be feasible. The shaft positions and the corresponding base gear ratios are derived for each of them. As an example of the novel type of ICT, a transmission is going to be presented with the PGTs shaft positions as in Fig. 2. The mutually constant speed ratios between PGT shafts are defined as:

$$\alpha = \frac{n_3}{n_2} = \frac{z_2}{z_3} > 0 \tag{1}$$

$$\gamma = \frac{n_5}{n_4} = \frac{z_4}{z_5} > 0 \tag{2}$$

$$\beta = \frac{n_{cm}}{n_{out}} > 0 \tag{3}$$

$$\delta = \frac{n_{10}}{n_2} = -\alpha i_G \frac{z_3}{z_1} \tag{4}$$

$$\varepsilon = \frac{n_6}{n_4} = -\frac{z_4}{z_6} = -\gamma \frac{z_5}{z_6} \tag{5}$$

$$\varphi = \frac{n_{12}}{n_{out}} = -\beta i_H \frac{z_8}{z_9}, \tag{6}$$

where n_i ($i = 1$ to 12) are the rotational speeds of gears 1 to 12, z_i are the number of teeth of gears 1 to 12 and $i_G = z_{11}/z_{10}$ and $i_H = z_{13}/z_{12}$ are the gear ratios of gear drives G and H, respectively.

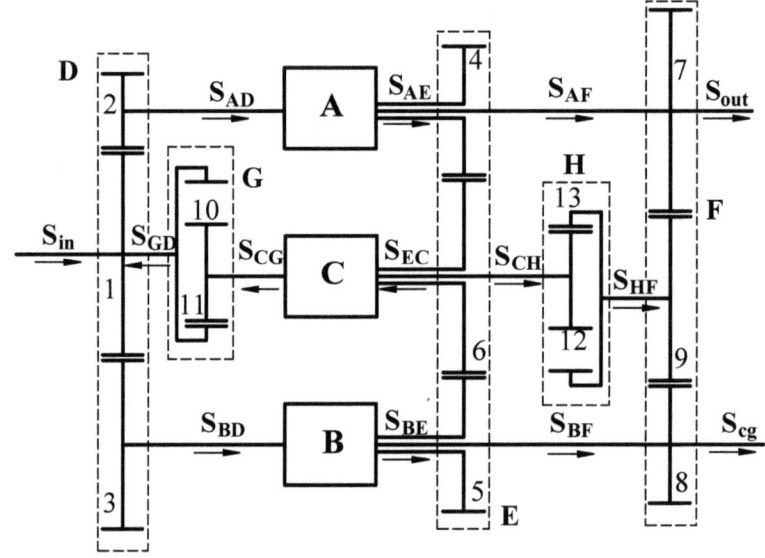

Fig. 1. Schematic illustration of the proposed transmission

The base gear ratio of PGT C is derived:

$$i_C = \frac{\delta - \varphi}{\varepsilon - \varphi}.$$ (7)

In regard to connections between PGTs A and C, the base gear ratio of PGT A should be:

$$i_A = \frac{\varepsilon}{\delta} \frac{\delta - \varphi}{\varepsilon - \varphi}.$$ (8)

Since it is to be equal to the already determined base gear ratio, Eq. (11), the conditions should be fulfilled:

$$\frac{i_G}{i_H} = \frac{z_1}{z_3} \frac{z_5}{z_6} \; ; \; \frac{z_6}{z_5} = \frac{z_9}{z_8}.$$ (9)

The feasibility condition for PGT C is found to be $\varphi < \delta < \varepsilon$ or $\varepsilon < \delta < \varphi$ which in accordance with Eqs. (4) to (9) converts to the feasibility condition for link between PGTs A and B: $\beta < \alpha < \gamma$ or $\gamma < \alpha < \beta$. This means there is no additional feasibility condition for PGT C.

3 NUMERICAL EXAMPLE

For the transmission shown in Fig. 2, the speed ratios are chosen: $\alpha = 10/9$, $\beta = 1$ and $\gamma = 2$ which result in the base gear ratios: $i_A = 1/5$; $i_B = 1/9$.

From Eqs. (1)-(6), (29) and (30), for obvious conditions of equality the centre distances between main axes of PGTs, for assumed equal modules of all gear drives and for chosen number of teeth $z_2 = 20$, $z_5 = 18$ and $z_7 = 20$, the number of teeth of all other gears are calculated, including those of PGTs, Table 1, and speed ratios: $\delta = -260/99$, $\varepsilon = -4/7$, $\varphi = -26/71$ are obtained. The base gear ratio of PGT C is then calculated as $i_C = 11$.

Gear mark	1	2	3	4	5	6	7
Number of teeth	71	20	18	36	18	63	20
Gear mark	8	9	10	11	12	13	1A
Number of teeth	20	70	18	26	48	39	16
Gear mark	2A	2'A	3A	1B	2B	2'B	3B
Number of teeth	53	16	53	45	15	45	15
Gear mark	1C	2C	2'C	3C			
Number of teeth	36	20	18	36			

Tab. 1. Number of teeth of the gears of example transmission

For three equal periods of time where in each of them the same block input rotational speed history of 200, 400 and 600 min^{-1} is given; the control generator speed is set to -600 for the first, -800 for the second and -1000 min^{-1} for the third period of time; the speeds of all shafts are calculated and those of the output and gear 4 shafts are drawn in diagrams in Fig. 6 together with input shaft speeds. The controllability of the transmission is clearly observed.

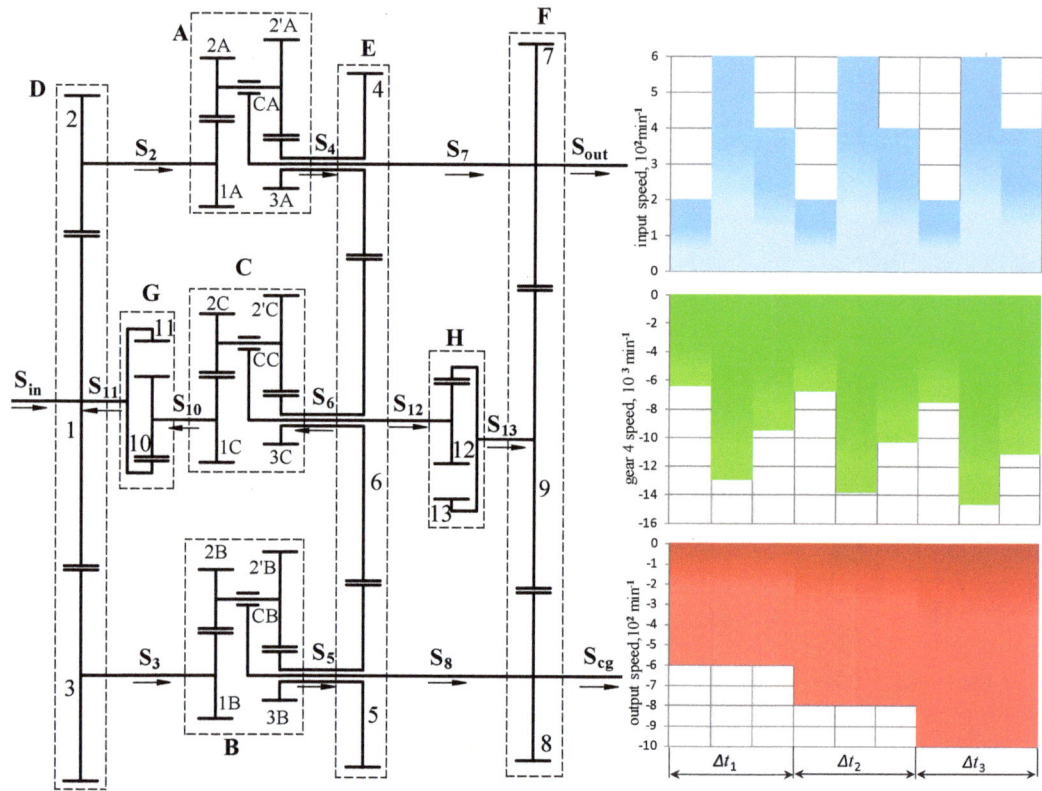

Fig. 2. Scheme of the proposed independently controllable transmission and diagrams of controlled output speed for assumed input speed history

4 CONCLUDING REMARKS

Kinematical analysis is presented of the so called independently controllable transmissions which are able to convert the variable or constant speed of transmission input shaft into the controllable speed of the output shaft set by will in the required range. The synthesis of those transmissions is carried out and its controllability is proved with a numerical example.

References:
[1]   Mantriota, G. *Theoretical and experimental study of a power split continuously variable transmission system, Part 1*. Proc. IMechE, Part D: J. Automob Eng, 2001, 215, 837-850.
[2]   Mantriota, G. *Theoretical and experimental study of a power split continuously variable transmission system, Part 2*. Proc. IMechE, Part D: J. Automob Eng, 2001, 215, 851-864.
[3]   Mucino, V. H. et al. *Design of continuously variable power split transmission for automotive applications*. Proc. IMechE, Part D: J. Automob Eng, 2001, 215, 469–478.
[4]   Srivastava, N. & Haque, I. U. *On the operating regime of a metal pushing V-belt CVT under steady state microslip conditions*. Proc. Intern Cont Variable and Hybrid Transm Congress, San Francisco, 23-25 September 2004.
[5]   Jamieson, P. *Innovation in Wind Turbine Design*, John Wiley & Sons, Chichester, UK, 2011.
[6]   Hwang, G. S. et al. (2011). *Kinematic Analyses of a Parallel-type Independently Controllable Transmission*. Int. J. Autom. Smart Technol., 1, 87-92.

PROPERTIES AND APPLICATION OF POLYMER CONCRETE IN DESIGN OF FRAMES PARTS OF MACHINE TOOLS

Limberg, L. & Lašová, V.

Abstract: This paper is focused on properties of polymer concrete and its application in design of machine tools. Basic information about composition of polymer concrete mixtures and their properties is presented. Methods of identification of material properties for purposes of calculations using FEM are described.

Keywords: polymer concrete, composite materials, property of materials, unconventional materials.

1 INTRODUCTION

Polymer concrete is increasingly used material that is alternative for cast-iron and steel in design of beds of machine tools. It is particle composite material composed of minerals which serve as a filler. The matrix consists of resin and a hardener. This material provides interesting and design-advantageous properties, e.g. damping, dimensional and thermal stability, high stiffness, corrosion resistance, etc.

2 COMPOSITION AND TECHNOLOGY OF PREPARATION OF POLYMER CONCRETE MIXTURES

The mixture for casting of polymer concrete casts is composed of two basic components – filler and matrix. These components are modified during the technology procedure and are mixed in the final phase, thus the casting final mixture is created.

2.1 Fillers
Examples of minerals, which serve as fillers, are granite, quartz or basalt. The selected filler and its structure have a major effect on the properties of the casting. Filler may be powder (e.g. quartz, feldspar, calcite ...), fibres (e.g. glass, carbon) or pads (e.g. talc, mica). The determining factors for quality of the mixture are particular shape, size, distribution and the proportion of filler in the mixture.

2.2 Matrix
The matrix is a mixture of a resin and a hardener. There are various types of resins, e.g. polyester or polyurethane resin, but the most common resin used in the matrices is an epoxy resin. Its advantage is lower volumetric shrinkage in comparison with other types. Epoxy resin is also characterized by adhesion to the selected type of the bond and dimensional stability, which is key parameter in the in terms of design of machines working with high precision.

2.3 Additives

Various additives are added to polymer concrete in order to improve the quality of the finished cast by the ventilation, enhanced leaking into the mold, reduction of viscosity and by improved adhesion of the mixture. It also includes separation agents, which facilitate removing of the finished product from the mold.

2.4 Preparation of the mixture

Preparation of the mixture is divided into two main phases. In the first phase the fillers of different particle sizes are mixed. The resin and hardener are mixed separately. In the second phase both mixtures are mixed together using screws. Thanks to mixing these components, a final mixture with a higher viscosity is created, which allows an easier casting. The ratio of matrix and fillers and their exact composition is various according to required properties of final product. It is also a "know-how" protected by the manufacturers against competition. However, in general, the ratio between the filler and the matrix varies around 80:20. The final mixture can be casted directly from mixing unit into molds or into ladles.

3 SUMMARY OF IMPORTANT PROPERTIES OF POLYMER CONCRETE

Thermal stability – thanks to high thermal capacity in combination with low thermal conductivity the polymer concrete responds slower to temperature changes in its surroundings in comparison with cast-iron.

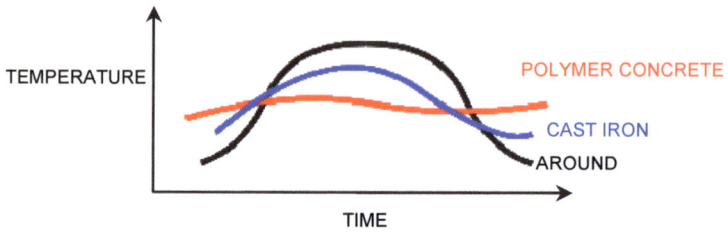

Fig. 1. Thermal stability of polymer concrete

Damping properties – as mentioned before, polymer concrete achieves up to 10-times better damping properties than cast iron has. Due to that fact, it became convenient material for design of machine tools, because it improves the precision of machine tools. Thanks to its properties, in comparison with cast-iron polymer concrete reduces of number of free oscillations and allows the shift of eigenfrequencies out of critical range.

Internal tension of the cast – after casting and subsequent cooling internal tensions occur, caused by shrinkage of the cast. These tensions are minimal due to the composition of the polymer concrete, specifically epoxy resin characterized by a low shrinkage. Good polymer concretes achieve shrinkage in the range from 0,02 to 0,03%. As the manufacturers state, the dimensions are tolerated (within shrinkage) 0,3 mm to 1 m.

Degradation of polymer-concrete products – polymer concrete itself is a very durable material, but there are factors that can degrade the final product. These factors include the degradation of corrosion reinforcement, salt assault, acid and sulphate treatment, mechanical abrasion and more.

When comparing the mechanical properties of polymer concrete with metals, polymer concrete has specific advantages and disadvantages. One major advantage of concrete is its damping properties, which can be up to 10 times better than steel or cast iron. That is the reason why it is used in the design of high-precision machine tools. Another advantage is its density, which is 3 times lower, which makes final products lighter. The advantage of metal materials is especially their tensile, compressive and flexural strength, which reach up to 10 times higher values than polymer concrete can provide.

4 MECHANICAL PROPERTIES OF POLYMER CONCRETE

Considering the utility properties of the polymer concrete mentioned above, it is clear that its application to the fixed frame parts of machine tools has considerable perspective. Due to the lower strength of the material and the manufacturing method, it is clear that the frame parts of machines will be different from the conventional metal part. In frames of machine tools hollow prismatic shapes will be most applied. Nowadays, design of new machine parts is commonly accompanied by a variety of computational simulations and optimizations in order to design the piece with the best usage of the material. For parts made of alternative materials computational assessment of the final temperature – mechanical behaviour of the proposed component is necessary.

For the prediction of properties of new structures simulation using FEM are used. Their correctness is highly dependent on the accuracy of input data – in this case on the input material properties. When performing computational simulations, it is suitable not to rely on the values specified by the manufacturer, but to verify these values by measurement. Measurement procedures are defined by standards (DIN 51 290) – Tests of mineral iron in the construction of machine tools.

In the framework of the research project tests of mineral iron specimens were performed in order to verify the basic necessary values for calculation of the frame part: tensile modulus E and Poisson's ratio μ. These values are sufficient, as the mineral iron can be considered as isotropic material governed by Hooke's law.

Compression, bending and tensile tests were performed, each on five specimens. From the values obtained by measuring machine and data from strain gauges in the elastic region, the elastic moduli of specimens E and Poisson's ratio (as the ratio of transverse and longitudinal deformation) can be calculated.

During the compression test, interesting behaviour of the material was observed. In about 20% of the yield strength, the material can actually be considered as elastic. Moderate nonlinearity in initial growing is probably caused by the filling the gaps during loading. In the next step of loading (about 40% of the yield strength) delay occurs, which may be caused by creep of the matrix. The third stage shows hardening. Overall, the material behaves brittle.

Three-point bending test was performed on normalized specimens using tensile machine. The fracture surface of the specimen was perpendicular to its longitudinal axis and it was apparent on the fracture surfaces that the initiator of damage was greater particle of grit.

Also, this test can be used to verify the values of modulus of elasticity. Determined values of flexural modulus show some dispersion, which is understandable in case of heterogeneous material. Value of the modulus of elasticity E of checked specimens after statistical evaluation was 36 GPa.

The determined value of modulus of elasticity is low in comparison with steel or cast iron, so it is confirmed that the resultant frame part of the machine tool must be designed in a completely different way from conventional parts. On the other hand, since the frame parts of the machine tool are not subjected to high stresses, it is important to design a part with high stiffness. Experimental bed for large machine tool of mineral cast will be part of the project proposed in the next year.

5 CONCLUSIONS

Mineral casts can contribute significantly to improving the quality and increasing the useful properties of machine tools, which require high precision. Mechanical and thermal properties of the polymer concrete ensure more stable and quieter operation conditions of machining when compared with conventional materials.

While respecting the different properties of the mineral iron, the final form of the parts can be designed and computationally optimized. Due to the simplicity of technical training and undemanding finishing operations on the casting, remaining biggest disadvantage compared to cast iron and metal materials is its price.

Acknowledgements:
This work is supported by the Technology Agency of the Czech Republic, Project TE01010075 "Competence Center – Manufacturing Technology"

References:
[1] Jančář, J. *Úvod do materiálového inženýrství polymerních kompozitů*. Vyd. 1. Brno : Vysoké učení technické v Brně, Fakulta chemická, 2003, p. 193, ISBN 8021424435.
[2] *Schneeberger.com* [online]. 2011 [cit. 2011-05-24]. Mineral casting – Schneeberger. <*http://www.schneeberger.com/products/mineralcasting*>
[3] Daďourek, K. *Kompozitní materiály – druhy a jejich využití*, Technická univerzita v Liberci, 1. vydání, 2007, p. 113, ISBN 978-80-7372-279-1.
[4] Kosnar, M. & Lašová, V. *Závěrečná zpráva projektu 1.2.2*, Praha 2005.

KEY COMPARISONS OF NATIONAL ETALONS IN THE FIELD OF AMPLITUDE AND SPACING TEXTURE PARAMETER MEASUREMENTS AT THE NANOSCALE

Lyssenko, V.; Rakitin, J. & Bogomolov, D.

Abstract: The results of key comparisons of national Russian and Belarus etalons in the field of amplitude and spacing surface texture parameter measurements at the nanoscale are described. The etalon measure with nominal for parameters R_{max} and S_m was used as a standard. The real meaning of amplitude and spacing parameters were defined. The degree of equivalence of national etalons in the field of amplitude and spacing parameter measurement was defined.

Keywords: *national etalons, key comparisons, roundness, measurement system.*

1 INTRODUCTION

International organization for metrology – COOMET organized key comparisons of national etalons in the field of amplitude and spacing surface texture parameter measurements. Procedure of key comparisons of national etalons in the field of amplitude and spacing surface texture parameter measurements includes:
- definition of measuring procedure,
- working out of mathematical model for parameter measuring procedure,
- definition and calculating of uncertainty of parameter measurement budget,
- working out of mathematical model for parameter measuring procedure,
- definition and calculating of uncertainty of amplitude and spacing surface texture parameter measurement budget,
- definition of the degree etalon equivalence.

In 2013 year the key comparisons of Russian and Ukrainian national etalons in the field of height and lateral length parameter measurements was executed [1]. Present work shows the results of the key comparisons of Russian and Belarus national etalons in the field of amplitude and spacing surface texture parameter measurements. The key comparisons were performed in the Russian Research Institute for Metrological Service of Russia.

2 METHODS FOR MEASURING THE AMPLITUDE AND SPACING PARAMETER AT THE NANOSCALE

For the selection the best method for creating national etalons were executed comparisons of different methods of amplitude and spacing surface texture parameter measurements at the nanoscale [2].

In Atomic force microscopy (AFM) the tip with radius of curvature commensurable with the atomic size in located near to a surface in a scope of atomic forces. Force between probe tip and a surface is used to determine mechanical properties of a surface of researched object and its geometry. Such devices allow to receive surface images with the horizontal and vertical resolution up to 0,1 nm that is their basic advantage before optical and contact methods.

The next device the new modification of interference microscope MII-4 was developed in VNIIMS (Russia) for the automatic identification of the achromatic land in white light interferometers and for the digital measurements of its centre position. In interference microscope MII-4 the wavelength is used as standard (it is the natural standard). Therefore this interferometer can be used as primary standard for traceability in nanometer range.

Talystep is a stylus instrument designed for topography at the atomic to micron level, which has a vertical resolution better than 1,0 nanometer. It provides simple, direct measurement of nanotopography, and produces permanent graphical recordings of step height and surface texture, magnified up to 2 000 000 times. We used this devices for investigation its ability for precision measurements of amplitude and spacing surface texture parameters. The results of investigations were compared with methods of SPM and interference microscopy in this kind of measurements.

Both VNIIMS (Russia) and Belgim (Belarus) selected method of contact profilometry for creating national etalon.

3 TECHNICAL DESCRIPTION OF THE RUSSIAN NATIONAL ETALON IN THE FIELD OF SURFACE TEXTURE PARAMETER MEASUREMENTS AT THE NANOSCALE

Measurement system consists of high precision profilometer Talystep, analogue-digital converter (ADC/DAC) board LCard, special made electronic board, connection cables, and personal computer with installed software [2].

The central element of this system is Talystep profilograph, which includes measuring colon, driving gear with remote control, stylus measuring gauge, and analogue control device with plotter. During the measurement process Talystep allows to set zoom scale, stylus speed (mechanically), and stylus movement direction.

Measuring gauge translates profile heights into analogue signal that is transferred to the electronic control unit and at least to the plotter. In the measurement system the Talystep profilograph is connected to analogue input ADC/DAC board by means of interface cables. ADC/DAC board is connected to PC by means of USB interface. Special made electronic board for motor give control is connected to ADC/DAC digital output and profilometer drive control unit.

During measurement process stylus gauge transforms surface profile height into electrical signal which is then proceeded to ADC. Obtained digital signal from ADC/DAC board is transferred to PC by means of USB-interface and received by Msiu TalyStep software that processes digital analysis of measurements results. Stylus drive movement is controlled by MsiuTalyStep software. Drive start/stop signals are transferred through ADC/DAC board to special made interface board and then to profilometer drive control.

Talystep software provides both preliminary measurements of single gauge vaPlues, which are immediately displayed in gauge window, and full-processed profile measurement with sophisticated analysis. During profile measurement procedure customer can choose desired magnification, drive direction and velocity, horizontal discretization. Measurement could be done with static (immovable) drive unit that allows to assess measurement error caused by mechanical vibrations and electrical noise. Example of the measurement of sample amplitude and spacing surface texture parameters with 3 closely located deep marks is presented in Fig. 1.

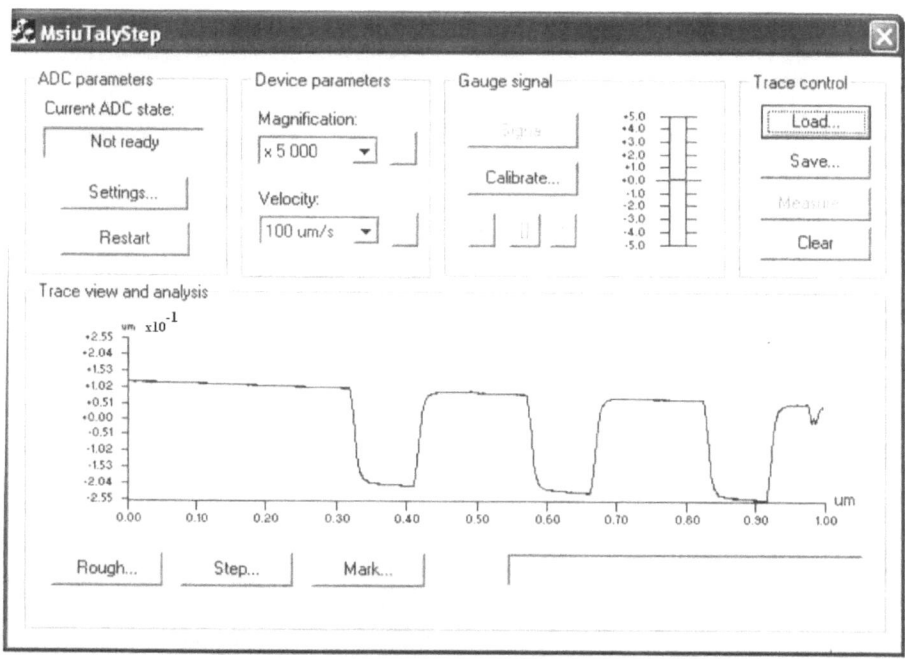

Fig. 1. Sample etalon measurement with 3 marks

4 PROCEDURE OF KEY COMPARISONS OF NATIONAL ETALONS

The procedure of comparisons includes [1]:
- working out of mathematical model for amplitude parameter measuring procedure,
- definition and calculating of uncertainty of amplitude parameter measurement budget;
- working out of mathematical model for spacing parameter measuring procedure,
- definition and calculating of uncertainty of spacing parameter measurement budget,
- definition of the degree etalon equivalence.

Mathematical model for spacing surface texture parameter S_m measurement is:

$$S_m = S_{m_M} + \delta_c + \delta_{Rq} + \delta_M + \delta_o, \tag{1}$$

where S_{m_M} is the measured meaning of the parameter S_m etalon, δ_c is the type B uncertainty of the measuring system, δ_M is the type B uncertainty of etalon measure, δ_o is the type B uncertainty of measuring system resolution. Standard uncertainty $u(S_m)$ of the measuring S_m value is:

$$u(S_m) = \sqrt{c_{S_{m_u}}^2 \cdot u^2(S_{mu}) + c_{\delta_c}^2 \cdot u^2(\delta_c) + c_{\delta_{Rq}}^2 \cdot u^2(\delta_{Rq}) + c_{\delta_m}^2 \cdot u^2(\delta_m) + c_{\delta_o}^2 \cdot u^2(\delta_o)}, \tag{2}$$

where $c_{Ra_u} = c_{\delta_c} = c_{\delta_{Rq}} = c_{\delta_m} = c_{\delta_o} = 1$. The expanded uncertainty of the spacing parameter S_m

measurement is:

$$U(S_m) = 2 \cdot u(S_m). \tag{3}$$

Mathematical model of amplitude surface texture parameter $R_{\max}$ measurement is:

$$R_{\max} = R_{\max M} + \delta_c + \delta_{Rq} + \delta_M + \delta_o, \tag{4}$$

where $R_{\max M}$ is the measured meaning of amplitude surface texture parameter $R_{\max}$ etalon. Standard uncertainty $u(R_{\max})$ of measuring $R_{\max}$ value is:

$$u(R_{\max}) = \sqrt{c_{R\max_u}^2 \cdot u^2(R_{\max_M}) + c_{\delta_c}^2 \cdot u^2(\delta_c) + c_{\delta_{Rq}}^2 \cdot u^2(\delta_{Rq}) + c_{\delta_m}^2 \cdot u^2(\delta_m) + c_{\delta_o}^2 \cdot u^2(\delta_o)}, \tag{5}$$

where $c_{R\max_u} = c_{\delta_c} = c_{\delta_{Rq}} = c_{\delta_m} = c_{\delta_o} = 1$. Expanded uncertainty of amplitude surface texture parameter $R_{\max}$ measurement is:

$$U(R_{\max}) = 2 \cdot u(R_{\max}). \tag{6}$$

On the base of investigating ability of amplitude and spacing surface texture parameter measurements the key comparisons of national standards of Belarus and Russia were executed. Results of comparisons showed a good reproducibility. On the base of executed research of the accuracy ability of surface texture measuring devices at the nanoscale the calibrating hierarchy scheme was created, which allows to achieve traceability of length measurements in nanotechnology. Results of the key comparisons of national etalons are shown in the table.

Nominal meaning of parameter, μm	Results of measurements Russia R_1, μm	Results of measurements Belarus R_2, μm	Difference in the results $|d|$, μm	Uncertainty of measurements Russia u_{c1}, μm	Uncertainty of measurement Belarus u_{c2}, μm
$S_m = 0{,}8$	0,822	0,820	0,002	0,007	0,006
$R_{\max} = 0{,}3$	0,309	0,312	0,003	0,014	0,017

Tab. 1. Results of the key comparisons

5 CONCLUSION

Results of key comparisons of national etalons in the field of amplitude and spacing surface texture parameter measurements showed, that national etalons of Russia and Belarus in this field of amplitude and spacing surface texture parameter measurements are equivalent.

References:
[1] Lyssenko, V. & Kononogov, S. *Supplementary comparisons of national etalons in the field of surface roughness measurements.* Proc. of the 8th International Conference Cadam 2010.
[2] Poroshin, V.; Bogonlov, D.; Lyssenko, V. & Kononogov, S. (2008). *High Precision PC Based Measurements System for etalon Roughness analysis.* Advanced Engineering Vol. 2 No. 2, ISSN 1846-5900.
[3] Loukyanov, V. S. & Lyssenko, V. G. (1991). *Accuracy of determination of microtopographical parameters discrete and analog methods*, Measuring engineering, №9.

AN APPLICABLE SHORT-TERM TRAFFIC FLOW FORECASTING METHOD BASED ON CHAOTIC THEORY

L. Mrgole, A. & Sever, D.

Abstract: Short-term traffic flow forecasting plays a very important role in urban traffic management and control. In this paper, according to the chaotic property of urban traffic flow, we compute the parameters of phase space reconstruction for a traffic flow system. Meanwhile, a local-forecasting method is introduced to predict urban road short-term traffic flow based on the theory of phase space reconstruction.

A case study using real traffic flow data from real time traffic counter proves the validity of the method. The research in this paper is a significant attempt to forecast traffic flow from the viewpoint of non-linear time series.

Keywords: traffic, forecasting traffic flow, prediction, chaotic parameters.

1 INTRODUCTION

Application of Intelligent Transportation Systems (ITS) gave the researchers the ability to research more deeply on traffic flow analysing and processing. Traffic flow is the most important and fundamental information for all its subsystems such as Advanced Traffic, Management System, Advanced Travel Information System and Advanced Public Transportation System. Furthermore, only if the traffic flow information is obtained in time and accurately, can the effectiveness of traffic control and management be evaluated.

Literature has shown that traffic flow time series has some chaotic properties, which proves that it is applicable to study traffic flow using nonlinear theory. It analyses the auto-similarity during the change process of traffic flow, which proves the predictability of traffic flow. All these literatures provide a new approach for the short-term traffic flow forecasting.

This paper applies phase space reconstruction theory to forecast short-term traffic flow from the nonlinear time series point of view. Some satisfactory results have been achieved. This paper introduces some basic knowledge on phase space reconstruction. Furthermore, how to select the phase space reconstruction parameters including delay time and embedding dimension is also discussed. Moreover, traffic flow forecasting method and corresponding theoretical analysis based on phase space reconstruction are presented. Finally, a case study is introduced. The short-term traffic flow is predicted with real data obtained from traffic counters.

2 FORECASTING MODEL OF TRAFFIC FLOW BASED ON CHAOTIC NEURAL NETWORK

As we mentioned above, traffic flow system is a non-linear system with chaotic features. Traffic flow system appears to be irregular, but it still has inherent certainty. For example, at a certain period of time and at a certain part of the road, when quantity of cars reaches its overcapacity limit, the movement behaviour of a single car might occur as a random and irregular one. But, the driver's destination and time of arrival are clearly certain. It is exactly because of this, that chaotic phenomena appear and that some of the chaotic features can be predicted. BP neural network, because of its learning and training characteristics, always needs large amounts of samples and in reality it is hard to get by so much material. So, we will use chaotic features of traffic flow and by joining time-sequence, previously phase space reconstructed with neural network, make predictions on traffic flow.

This method of identification of chaotic patterns in traffic flow combines processing of time-sequence, theory of phase space reconstruction and neural network predictions in order to predict traffic flow within a period of time. The main idea is to: first, determine chaotic features in time-sequence, then use theory of phase space reconstruction to reconstruct phase space and finally after reconstruction, enter gathered data into BP neural network as to do exercises and consequently make predictions.

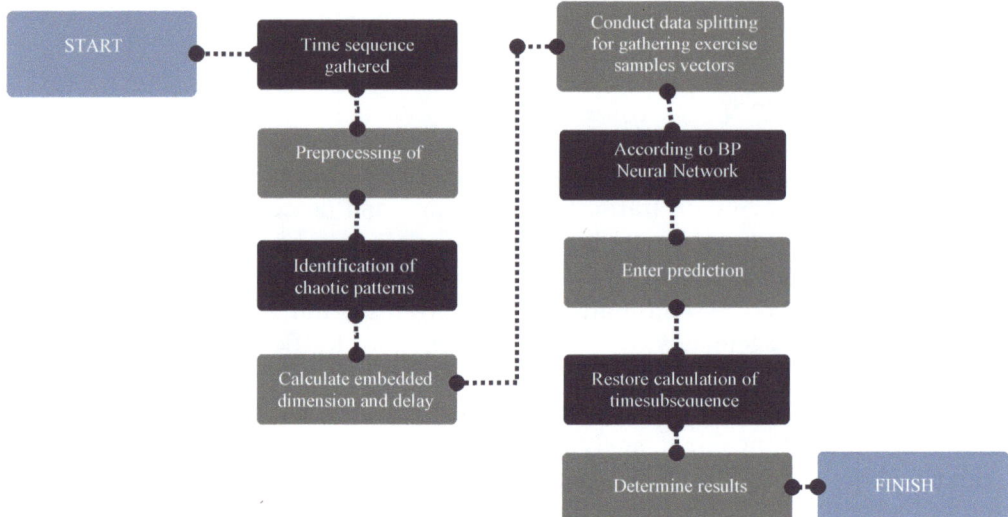

Fig. 1. Process of the model prediction of traffic flow

The main steps of this method are as follows:

Step 1: In order to make chaotic features found in observing traffic quantity in time sequence more clear and more suitable for neural network, we need to make pre-processing of the traffic quantity time-sequence, and this pre-processing should contain smoothing and differential method.

Step 2: As for pre-processed time-sequences, if chaotic neural network model is used for the first time to make predictions, you must do chaotic pattern identification on processed time-sequence to see if chaos theory can be applied.

Step 3: Phase space reconstruction must be done on traffic flow time-sequence, which corresponds to chaotic features in order to restore its attractor features. After reconstruction, this sequence will transform from a vector into m dimensional matrix.

Step 4: This step is crucial for combination of chaos theory and neural network. Because of phase space reconstruction, vectors cannot be immediately entered into neural network for exercises and predictions, but instead they have to be split and then entered.

For short introduction on main idea in data splitting, please refer to the following section.

Step 5: Exercise vectors which we acquired by splitting data in a way, which is described in step four enter into network in order to make exercises. After you find optimal approximation of F, and use this F to enter prediction vector to get prediction value.

Step 6: Do reverse calculations on the data gathered in step 5 and retrieve real predictive value of original time-sequence.

3 CASE STUDY AND RESULTS

We use the traffic flow data from the system real time traffic counters. The sampling time interval is 15min. We have done some experiments and our method has shown good performance. Here, we just randomly select one of them to show our results. The raw data from Mar. 1 to Mar. 21, 2013.

The raw data is shown in Fig. 3. The data of 20 days, as training set, are used to construct initial phase space. The residual data of 2 days, called testing set, are used to test the accuracy of the model proposed in this paper. Use chaotic neural network and BP neural network to make predictions on traffic flow time-sequence. Their results and real results are compared below:

	operation time	number of steps	average deviation	polarity
Conventional BP neural Network prediction	8,2456345	1560	0,0082542	0,87112
Chaos theory and BP neural network prediction	4,314568	712	0,0000451	0,9987

Tab. 1. Comparison of two prediction methods

Table 1 shows that prediction method based on chaos theory and BP neural network is slightly more precise, its simulation features are relatively better, calculation time is also relatively shorter, which makes it more apt for real-time control and guidance of traffic flow.

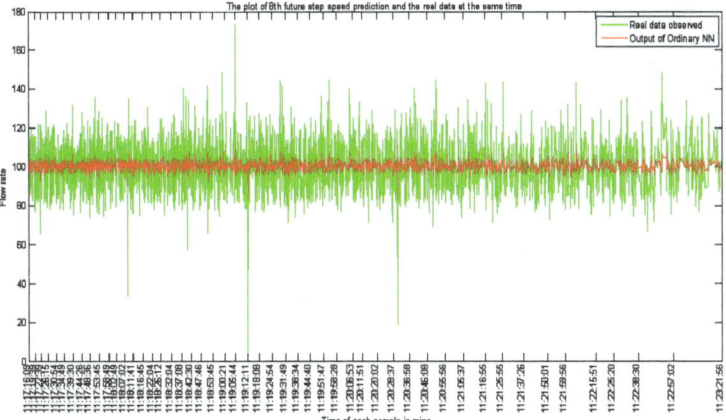

Fig. 2. Real data vs chaotic neural network

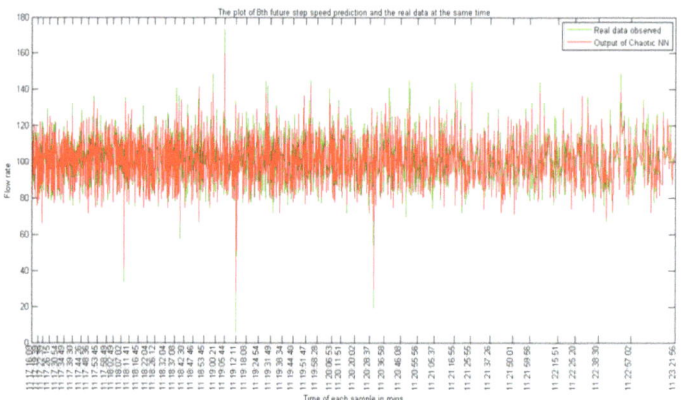

Fig. 3. Real data vs modified chaotic neural network

The prediction method based on chaos theory and BP neural network is slightly more precise, its simulation features are relatively better, calculation time is also relatively shorter, which makes it more optimal for real-time control and guidance of traffic flow.

4 CONCLUSION

Based on chaotic property of traffic flow time series, this paper introduces phase space reconstruction theory to forecast short-term traffic flow. The parameters used in phased space reconstruction such as delay time and embedding dimension are computed. Local linear regression is also introduced to seek near neighbour. Case study shows the method proposed is effective and applicable. This paper did some beneficial attempts in short term traffic flow forecasting from the aspect of non-linear time series analysis.

Through emulate results, the means of chaos neural network outlook is better than traditional neural network outlook in running time of algorithm, prediction accuracy and real-time. The prediction accuracy will definitely enhance provided improvement in the length of time series.

References:
[1]   Franklin, R. E. (1961). *The Structure of a Traffic Shock Wave.* Civil Engineering Pulb. Wks. Rev. 56, pp. 1186-1188.
[2]   Del Castillo, J. M. (1996). *A Car-Following Model based on the Lighthill-Whitham Theory.* In: Lesort, J. B. (ed), Proceedings of the 13th International Symposium of Transportation and Traffic Theory, Lyon, pp. 517-538.
[3]   Hilborn, R. C. *Chaos and nonlinear dynamics: an introduction for scientists and engineers.* 2nd edition. Oxford: Oxford University Press, 2001.
[4]   Casdagli, M. (1989.) *Nonlinear prediction of chaotic time series.* Physica D 35, pp. 335-56.
[5]   Lorenz, E. N. *The essence of chaos.* University of Washington Press, Seattle, 1993.
[6]   Disbro J. E. & Frame, M. *Traffic Flow Theory and Chaotic Behavior.* Transportation Research Record 1225, TRB, National Research Council, Washington DC, 1989. p. 109.
[7]   Bellomo, N. & Ridolfi, L. (1995). *Solution of nonlinear initial-boundary value problems by Sine collocation methods,* Computers Math. Applic. 29 (4), pp. 15-28.
[8]   Kaya, S.; Kilic, N.; Kocak, T. & Gungor, C. (2014). *From Asia to Europe: Short-Term Traffic Flow Prediction between Continents.* 21st International Conference on Telecommunications (ICT), pp. 277-82.

HIGH PRECISION PC BASED MEASUREMENT SYSTEM FOR 3D ANALYSIS OF SURFACE TEXTURE AT THE NANOSCALE

Poroshin, V.; Bogomolov, D.; Poroshin, O. & Lyssenko, V.

Abstract: The PC based measurement system for high precision 3D analysis of surface topography at the nanoscale is described. The system is based on the atomic-force metrological microscope with modified specimen table having extended horizontal range. System implies the 3D surface texture analysis according to the recent standard ISO 25178-2:2012.

Keywords: surface topography, nanoscale, measurement system, automation.

1 INTRODUCTION

Precision measurements of surface texture at the nanoscale are extremely required in many branches of modern precision engineering such as production of laser gyroscopes, laser mirrors, night vision devices, microchips etc. The atomic force microscopy is commonly used for the analysis of nanostructures on the surface.

Most of existing microscopes involves mainly qualitative assessment with little quantitative facilities. At the same time, the detailed parametrical quantitative technique of the surface texture analysis is well known and widely used at the microscale. The same analysis technique can be carried at the nanoscale as well.

Existing surface texture analysis standards includes traditional 2D surface profile analysis by ISO 4287-98 and recently involved 3D topography analysis by ISO 25178-2:2012. Precision surface texture measurements at the nanoscale are naturally intended to be analyzed by more accurate and informative 3D technique.

Present article describes the development of high precision PC based measurement systems based on metrological atomic force microscope NanoScan 3Di. System implies the 3D surface texture analysis according to recent standard ISO 25178-2:2012.

2 MEASUREMET SYSTEM STRUCTURE

The main technical properties of the proposed measurement system are presented in Tab. 1. Photo of the measurement system is presented in Fig. 1.

Measuring system has a module architecture that is shown in Fig. 2. It consists of the atomic-force microscope, the high range coordinate table and a personal computer with controlling software and analytical software.

The central element of this system is the atomic force microscope NanoScan 3Di having a piezo-resonance probe with stiff console and solid mechanical indenters made of artificial diamonds.

Property	Description
Measurement principle	Atomic force microscopy
Probe type	Diamond indenter
Horizontal range (um)	500
Vertical range (um)	50
Horizontal resolution (nm)	< 0,1
Vertical resolution (nm)	< 0,1
Time resolution of measurements (ms)	1
Max. measurement speed (um/s)	30
Axis orthogonality error (rad)	0,01
Digital filtering	3D Gaussian
Measured parameters	Sa, Sq, Sp, Sv, Sz, Ssk, Sku, Sdq, Sdr, Sal, Str, Std, Smr, Sdc, Sxp, Vmp, Vmc, Vvv, Vvc, Spd, Spc, S5p, S5v, S10z, Sda, Sha, Sdv, Shv

Tab. 1. Main properties of the measurement systems

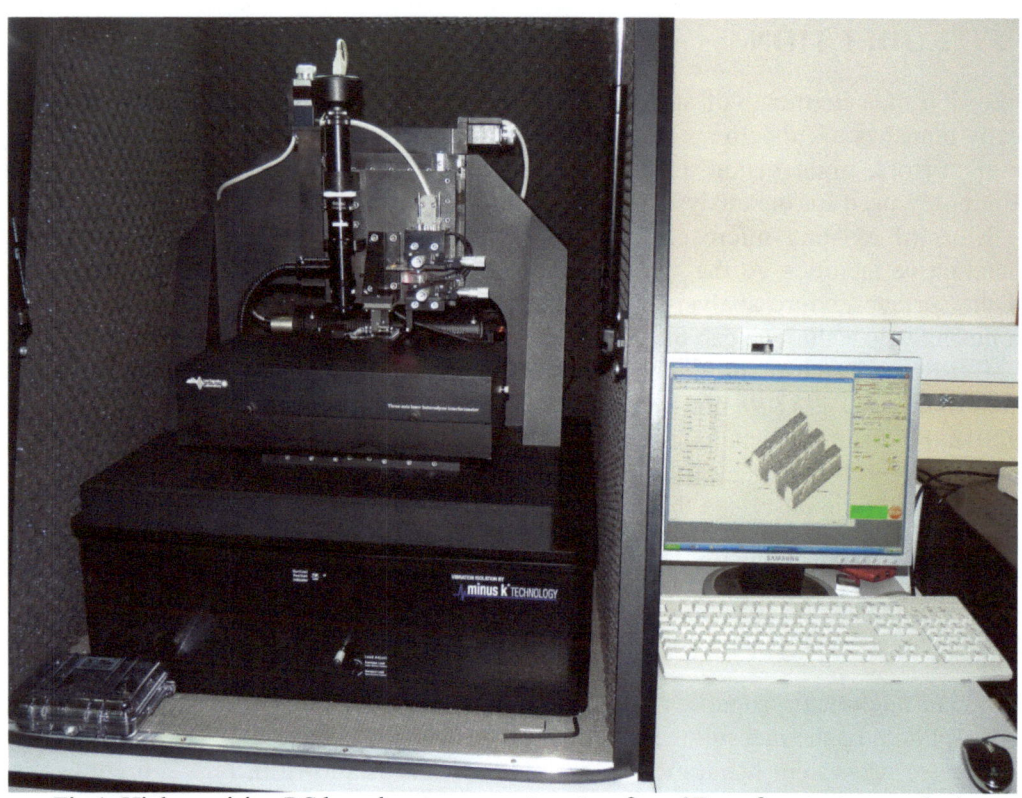

Fig 1. High precision PC based measurement system for a 3D surface texture analysis
at the nanoscale

Scanning of the surface texture implies horizontal movement of the specimen by means of three-coordinate table. Parametrical analysis of nanostructure texture imposes extended requirement to the horizontal range of the table. That is why system includes specially constructed high range and high precision three-coordinate piezo-table equipped with heterodyne laser interferometer having digital phase detector.

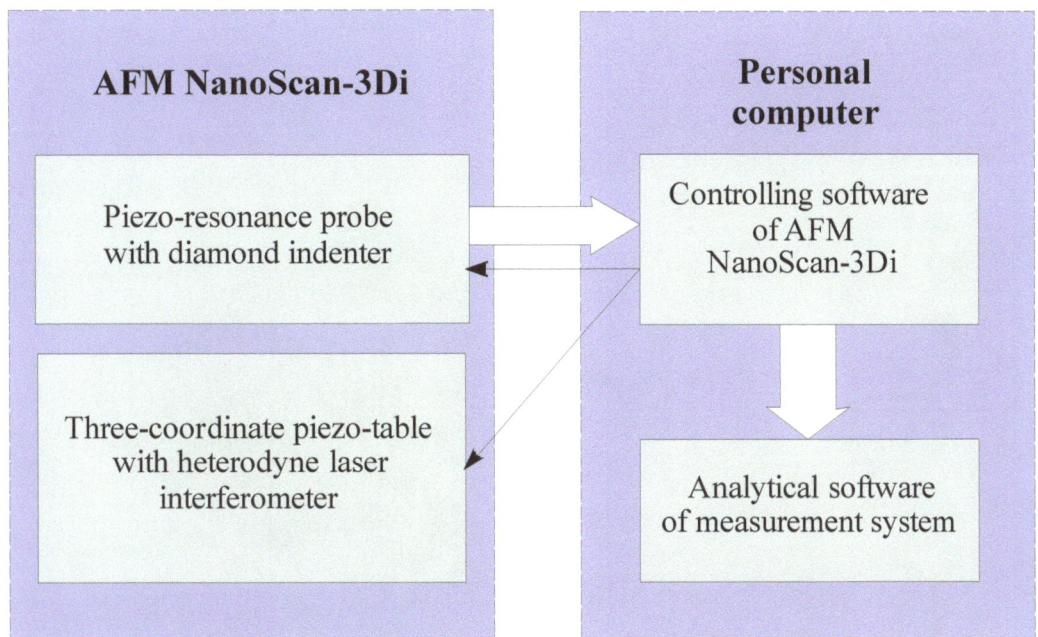

Fig. 2. Architecture of the PC based measurement system

Electronic module of the three-coordinate table has time resolution of 1 ms that allows to perform surface scanning at up to 30 μm/s speed. Measured data are transferred to the PC by means of USB interface.

Controlling software provides preliminary probe placement, motion control of table and probe, performing of single point measurement and incorporation of measured data into unified digital field.

After the measurement process is finished measured data are transferred to the analytical software. Analytical software implements surface visualization, form filtering (including specimen slope leveling), frequency filtering by means of digital phase-corrected Gaussian filter, parametrical assessment, correlation analysis and surface curve analysis. Parametric analysis includes 28 parameters that are divides into groups of amplitude and hybrid parameters, spatial parameters, functional parameters, and segmentation parameters.

Examples of the 3D analysis of the amplitude surface texture parameters of the grained surface and diamond surface in analytical software are presented in Figs. 3,4.

3 CONCLUSION

The proposed measurement system provides automated high precision measurements and complex PC based parametrical 3D analysis of surface texture at nanoscale. It is recommended to use in modern nanotechnology and precision engineering laboratories.

The research was performed with the financial support of the Ministry of Education and Science of the Russian Federation for higher education institutions within the state job service.

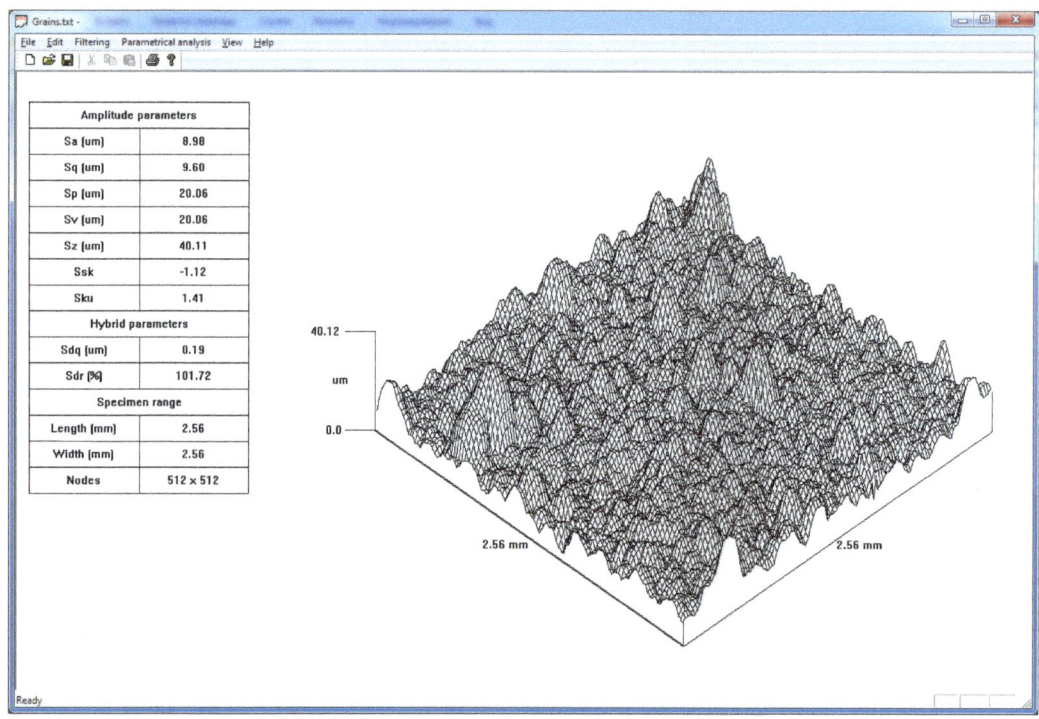

Fig. 3. Sample of 3D surface topography of the grained surface

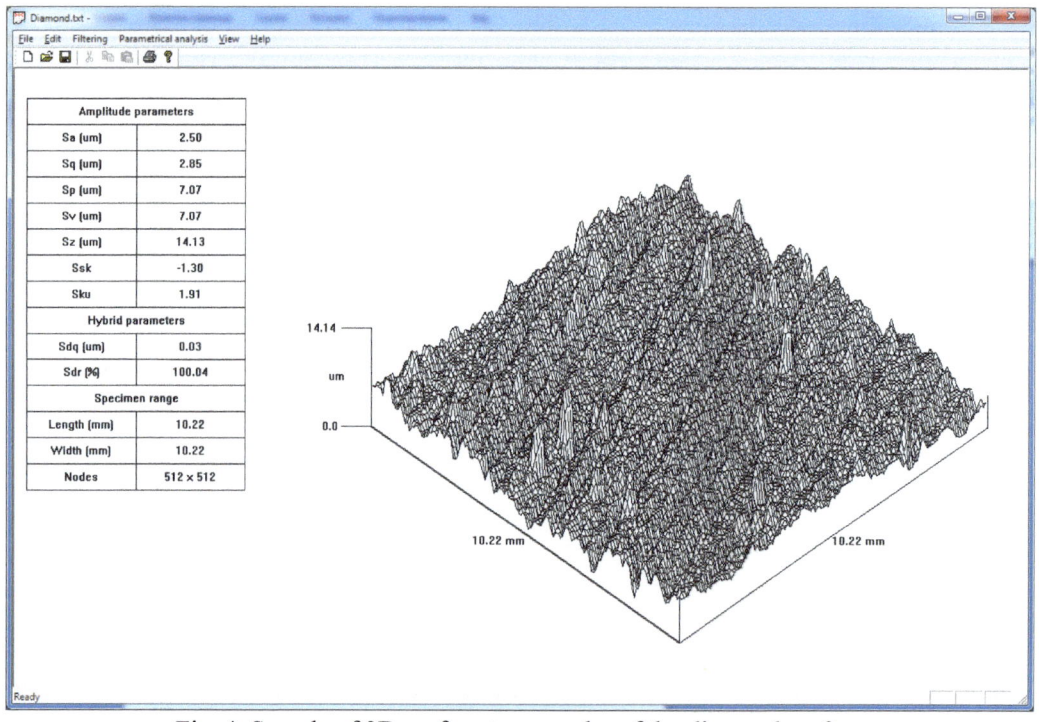

Fig. 4. Sample of 3D surface topography of the diamond surface

CONSOLIDATED RESULTS OF KEY COMPARISONS IN THE FIELD OF MEASUREMENT OF ROUNDNESS GEOMETRICAL PARAMETERS

Rakitin, J.; Lyssenko, V. & Bogomolov, D.

Abstract: The results of key comparisons of national etalons of Russia, Belarus and Ukraine in the field of roundness geometrical parameter measurements are described. Measurements of the roundness geometrical parameter were performed by means of PC based measurement systems Talyrond-73 (Russian etalon),Talyrond-565 (Belarusian etalon) and Talyrond-200 (Ukrainian etalon).

Keywords: *national etalons, key comparisons, roundness, measurement system.*

1 INTRODUCTION

In 2010 year the supplementary comparisons of national etalons Russia and Ukraine in the field of roundness measurements was executed [1]. In 2012 year the key comparisons of national etalons Russia and Belarus in the field of roundness geometrical parameter measurements was executed [2]. In this year the consolidated key comparisons of Russian, Ukrainian and Belarus national etalons in the field of roundness geometrical parameter was executed at the Russian Research Institute for Metrological Service of Russia. Present article describes the results of the key comparisons.

2 DEFINITION OF COMPARISONS MEASURING PROCEDURE

Procedure of key comparisons of national etalons in the field of roundness geometrical parameters measurements includes [3]:
1) working out the mathematical model for roundness geometrical parameters (parameter $P+V$) measuring procedure;
2) executing of key comparisons measurements on national etalons
3) definition of measured value and calculating of roundness geometrical parameters (parameter $P+V$) measurement budget uncertainty;
4) definition of the degree etalon equivalence.

3 THE MATHEMATICAL MODEL FOR ROUNDNESS GEOMETRICAL PARAMETER MEASUREMENT

The mathematical model of roundness geometrical parameters (parameter $P+V$) measurement is:

$$P+V = (P+V)e + \delta_c + \delta_{Rq} + \delta_M + \delta_o, \tag{1}$$

where $(P+V)e$ is the measured meaning of roundness geometrical parameter $P+V$ etalon, δ_c is the type B uncertainty of measuring system, δ_M is the type B uncertainty of etalon measure, δ_{Rq} is the type A uncertainty of measuring system RMS, δ_{o} is the type B uncertainty of measuring system resolution.

4 DEFINITION OF MEASURED VALUE AND CALCULATING OF ROUNDNESS GEOMETRICAL PARAMETERS MEASUREMENT BUDGET UNCERTAITY

Standard uncertainty u of $P+V$ measurement is:

$$u(P+V)=\sqrt{c_{\mathrm{Ra_u}}^2 u^2(P+V)e + c_{\delta_c}^2 u^2(\delta_c) + c_{\delta_{\mathrm{Rq}}}^2 u^2(\delta_{\mathrm{Rq}}) + c_{\delta_m}^2 u^2(\delta_m) + c_{\delta_o}^2 u^2(\delta_o)}, \qquad (2)$$

where the coefficients $c_{\mathrm{Ra_u}} = c_{\delta_c} = c_{\delta_{\mathrm{Rq}}} = c_{\delta_m} = c_{\delta_o} = 1$. The expanded uncertainty U of $P+V$ measurement is:

$$U(P+V) = 2\,u(P+V). \qquad (3)$$

Budget of parameter $P+V$ measurement uncertainty is shown in Table 1.

Budget of roundness geometrical parameters $P+V$ uncertainty measurement $(\times 10^{-2}\,\mathrm{m})$				
$P+V$	δ_c	δ_{Ra}	δ_m	δ_0
2,36	2,11	2,18	1,98	2,64

Tab. 1. Budget of roundness geometrical parameters $P+V$ uncertainty measurement

5 ETALON MEASUREMENT SYSTEMS

All three national etalons in the field of roundness geometrical parameters measurements are based on the high precision numerical roundness machines of the Talyrond series of Rank Taylor Hobson (UK) that are widely used in metrology services for verification and analysis of roundness and form geometrical parameters of rotation bodies. The detailed description of PC based measurement system based on Talyrond roundness machines is given in [4]. The national Russian roundness geometrical parameter measurement etalon is based on the roundness machine Talyrond-73 with rotating gauge. The national Belarusian roundness geometrical parameter measurement etalon is based on the roundness machine Talyrond-565 with rotating table. The national Ukrainian roundness geometrical parameter measurement etalon is based on the roundness machine Talyrond-200 with rotating table.

6 RESULTS OF KEY COMPARISONS

The key comparison of national etalons of Russia, Belarus and Ukraine was worked out in the Russian Research Institute for Metrological Service of Russia. The etalon measure of roundness geometrical parameters (parameters with nominal for $P+V = 12,0$ µm) was used as a standard for comparisons. The results of the etalon measurements are shown in Tables 2, 3, 4.

The final results of the calibration of etalon measures which was used in the further key comparisons of national etalons in the field of roundness geometrical parameter measurements are shown in Table 5.

No.	Measured Value $P+V$, (μm)	R.m.s.
1	12,53	
2	12,51	
3	12,58	
4	12,42	
5	12,51	
6	12,52	0,05
7	12,51	
8	12,57	
9	12,44	
10	12,51	
middle	12,510	

Tab. 2. Measurement results of roundness geometrical parameters etalon
(VNIIMS, Russia)

No.	Measured Value $P+V$, (μm)	R.m.s.
1	12,51	
2	12,55	
3	12,44	
4	12,52	
5	12,64	
6	12,58	0,06
7	12,53	
8	12,47	
9	12,51	
10	12,45	
middle	12,520	

Tab. 3. Measurement results of roundness geometrical parameters etalon
(Belgim, Belarus)

No.	Measured Value $P+V$, (μm)	R.m.s.
1	12,53	
2	12,55	
3	12,48	
4	12,54	
5	12,60	
6	12,57	0,06
7	12,55	
8	12,48	
9	12,55	
10	12,45	
middle	12,530	

Tab. 4. Measurement results of roundness geometrical parameters etalon
(Institute of Metrology, Ukraine)

Description	Value
Nominal meaning of parameter P+V) (μm)	12,5
Results of measuring in VNIIMS (Russia) (μm)	12,510
Results of measuring in Belgim (Belarus) (μm)	12,520
Results of measuring in Institute of metrology (Ukraine) (μm)	12,530
Difference of measurements between Russian and Belarus etalons (μm)	0,010
Difference of measurements between Russian and Ukrainian etalons (μm)	0,020
Difference of measurements between Belarus and Ukrainian etalons (μm)	0,010
Declared uncertainty of VNIIMS (Russia)(μm)	0,015
Declared uncertainty of Belgim (Belarus) (μm)	0,016
Declared uncertainty of Institute of metrology (Ukraine) (μm)	0,016

Tab. 5. Results of the key comparison of national etalons of Russia, Belarus and Ukraine in the field of roundness geometrical parameter measurements

The real meanings of the roundness geometrical parameters (parameter $P+V$) of all national etalons were defined during the key comparison procedure. The degree of the equivalence of national etalons in the field of roundness geometrical parameters (parameter $P+V$) was defined.

7 CONCLUSION

Key comparisons have shown that Russian, Belarus and Ukrainian national etalons in the field of roundness geometrical parameter measurements provide automated high precision measurements and complex PC based analysis of roundness geometrical parameters.

Results of key comparisons of national etalons in the field of roundness geometrical parameters measurements showed that national etalons of Russia, Belarus and Ukraine in the field of roundness geometrical parameters measurements are equivalent.

References:
[1] Poroshin, V.; Bogomolov, D; Kononogov, S. & Lyssenko, V. (2010). *Supplementary comparisons of modernized national etalons in the field of roundness measurements*, Advanced Engineering Vol. 4, No. 2.
[2] Kononogov, S.; Lyssenko, V. & Rakitin, J. (2013). *Key comparisons of national etalons in the field of height and lateral length parameters measurements in nanometer range*, Advanced Engineering, Vol. 7 No. 1.
[3] Loukyanov, V. S. & Lyssenko, V. G. (1991). *Accuracy of determination of microtopographical parameters discrete and analog methods*, Measuring engineering, No 9.
[4] Bogomolov, D.; Poroshin, V.; Kostyuk, A. & Radygin, V. (2010). *High precision PC based measurement systems for complex analysis of roundness and waviness*. Advanced Engineering, Vol. 4 No. 1, pp. 6-8.

THE ANALYTICAL FLOW RATE OF LINEAR PERISTALTIC PUMP CALCULATION METHOD

Sheypak, A. A.; Grishin, A. I. & Chicheryukin, V. N.

Abstract: The article presents the analytical method for calculation of linear peristaltic pump flow rate in the case of laminar flow. The comparison of the calculation results with the experimental data was carried out. It shows that the difference in the flow rate values decrease with the pressure.

Keywords: *mathematical model, peristaltic pump.*

1 INTRODUCTION

Peristaltic pump is a positive displacement pump with the flexible tube or hose element. A hose or tube element is positioned along the stationary pump housing and is compressed from the outside by a roller or another squeezing element. Peristaltic pumps are used in many applications – such as printing inks and colorings, wastewater slurries, mining slurries, bleach, food industry, beverages, chemical industry. Peristaltic pumps are also excellent for abrasive slurries and suction lift applications [1]. In most common design the roller moves along the outside of the element while the restitution of the hose or tube element behind the roller draws more fluid into the pump. Depending on the design of the pump the tube element may be placed bow-shaped, U-shaped, linearly or spirally. The peristaltic pump is often called linear if the tube element placed linearly and the pushers, which compress the tube in the same place in the cross-wise direction, are used as squeezing elements [2].

The tube deformation is often considered as the wave propagation. This approximation matches the peristaltic pump construction with bow-shaped, U-shaped, spirally placement of the tube and some constructions with linear placement of the tube. The fluid flow also often considered symmetrical. While a lot of different calculation methods presented for these constructions of the pump [3,4], few works related to the linear peristaltic pump exist. Mostly, these works include only the experimental research of certain construction of linear peristaltic pump. In this article the analytical method for calculation of linear peristaltic pump flow rate in the case of laminar flow is presented. The comparison with the experimental results shows that the difference in the flow rate values decreases with the pressure drop.

2 PROBLEM FORMULATION

The operation of the linear peristaltic pump with three steel plates used as squeezing elements may be described as follow. The first plate compresses the tube and act as valve sealing the tube. Next the second plate compresses the tube and pushes the operational fluid to the outlet of the pump. After that the first plate releases the tube and draws the fluid in the tube. Similarly, third plate squeezes the tube and then the first plate releases it. Finally, the second plate releases the tube and after that the third plate releases it.

3 THE ANALYTICAL METHOD OF FLOW RATE CALCULATION

Let us consider that the tube has length l and inner diameter d. Each plate compresses the tube during the time t_1 and holds it compressed during the time t_2. The time t_i is the interval between the plates action and t_0 is the total cycle time. The pressure at the inlet of the tube is p_1 and the pressure at the outlet of the tube is p_2. The flow rate Q can be found as

$$Q = \frac{t_0 - (t_1 + t_2 + 2 \cdot t_i)}{t_0} Q_{st} + \frac{t_1}{t_0} Q_1 + Q_{2,3}, \tag{1}$$

where Q_{st} – flow rate in the interval, when the tube is not compressed completely and plates doesn't move; Q_1 – flow rate in the interval, when the first plate compresses the tube; $Q_{2,3}$ – flow rate due to compression of the tube by the second and third plates. To find the flow rate Q_1 consider the fluid layer compression by two planes. The element of the upper plane with width $2a$ moves down with the velocity $V = h/t_1$, where h is the space between planes. The flow was considered laminar. For further analysis the lubricating layer approximation achieved by Reynolds in 1866 was used:

$$\frac{\partial p}{\partial x} = \mu \frac{\partial^2 v_x}{\partial y^2}; \; \frac{\partial p}{\partial y} = 0, \tag{2}$$

$$\frac{\partial v_x}{\partial x} + \frac{\partial v_y}{\partial y} = 0, \tag{3}$$

where μ is the dynamic viscosity of the fluid. Considering that no-slip condition takes place, boundary conditions are

$$y = 0, \; v_x = 0, \; v_y = 0; \; y = h, \; v_x = 0, \; v_y = -V; $$
$$x = -a, \; p = p_{a1}; \; x = +a, \; p = p_{a2}. \tag{4}$$

The pressure changes uniformly along the tube, therefore

$$p_{a1} - p_{a2} = 2 \cdot a \frac{p_1 - p_2}{l}. \tag{5}$$

After integrating the equation (2) one can obtain:

$$v_x = \frac{1}{2\mu} \frac{\partial p}{\partial x} y^2 + C_1 y + C_2 \tag{6}$$

Constants C_1 and C_2 can be defined by substitution of the boundary conditions (4). The result is

$$v_x = \frac{1}{2\mu} \frac{\partial p}{\partial x} y(y - h). \tag{7}$$

Next the equation (3) is used to define $\partial p / \partial x$. Multiplying equation (3) by dy and then integrating it yields

$$\frac{\partial}{\partial x} \int_0^h v_x dy - V = 0. \tag{8}$$

The substitution of expression (7) in the expression (8) results:

52

$$\frac{\partial^2 p}{\partial x^2} = -\frac{12\mu V}{h^3}.$$ (9)

Integrating the expression (9) and using the boundary conditions (4), we can obtain

$$\begin{cases} C_3 a - C_4 = -p_{a1} - \dfrac{6\mu V}{h^3}a^2; \\ C_3 a + C_4 = p_{a2} + \dfrac{6\mu V}{h^3}a^2. \end{cases}$$ (10)

The solution of the system (10) results:

$$\frac{\partial p}{\partial x} = \frac{p_{a2} - p_{a1}}{2a} - \frac{12\mu V}{h^3}x.$$ (11)

Substituting $\dfrac{\partial p}{\partial x}$ from (11) into (6) the result is

$$v_x = \frac{1}{2\mu}\left(\frac{p_{a2} - p_{a1}}{2a} - \frac{12\mu V}{h^3}x\right)y(y-h).$$ (12)

The flow rate by the unity of length is, substitution the expression (12)

$$q_1 = \int_0^h v_x dy, \quad q_1 = \int_0^h \frac{1}{2\mu}\left(\frac{p_{a2} - p_{a1}}{2a} - \frac{12\mu Va}{h^3}\right)y(y-h)dy.$$ (13)

To get the expression for flow rate Q_1 from flow rate by the unity of length q, we'll divide the expression (13) by h and multiply it by the cross section area S.

$$Q_1 = \left(\frac{p_{1a} - p_{2a}}{4\mu a} + \frac{6Va}{h^3}\right)\left(\frac{h^3}{6}\right)\frac{S}{h}, \quad Q_1 = \left(\frac{h^2(p_1 - p_2)}{12\mu l} + \frac{Va}{h}\right)S.$$ (14)

The flow rate $Q_{2,3}$ can be computed as the division of fluid volume, displaced by two planes, by the time of the whole cycle:

$$Q_{2,3} = (S \cdot 2a)\frac{2}{t_0}.$$ (15)

The flow rate due to the pressure difference between the inlet and outlet of the tube can be obtained from the Bernoulli's equation:

$$p_1 - p_2 = \rho(\alpha + \lambda\frac{l}{d})\frac{8Q_{st}^2}{\pi^2 d^4},$$

where λ – Darcy friction factor, α – kinetic energy correction factor. For laminar flow $\alpha = 2$ and $\lambda = 64/Re$, where Re is the Reynolds number, $Re = \dfrac{4Q_{st}\rho}{\pi d\mu}$, thus

$$p_1 - p_2 = \rho\frac{16Q_{st}^2}{\pi^2 d^4} + \frac{128\mu l Q_{st}}{\pi d^4}.$$ (16)

The solution of expression (16) is

$$Q_{st} = \frac{\pi d^2\left(\sqrt{\dfrac{256\mu^2 l^2}{d^4} + \rho(p_1 - p_2)} - \dfrac{16\mu l}{d^2}\right)}{4\rho}.$$ (17)

For occurrence of Q_{st} the cross section of the tube was considered circular thus the expression (20) is correct if $l \gg a$.

4 COMPARISON WITH THE EXPERIMENTAL DATA

The experimental facility included three plates driven by electromagnets which squeeze the flexible tube. Fluid (water) is pumped from the tank to the receiving vessel. The tank and the receiving vessel are placed on the different height. This difference creates the pressure drop $\Delta p = p_1 - p_2$ which was measured with accuracy 0,25 mm Wg. The tube length equals 1,5 m, tube inner diameter is 3 mm and tube outer diameter is 4,8 mm. Each plate is 14 mm width and space between plates is 4 mm. The maximum distance between plates and bump stop is 3,5 mm. Each plate compresses the tube during 0,016 s and keeps it compressed during 0,014 s. Time between the plates actuation is 0,02 s and the duration of whole cycle is 3 s. The pumped volume was measured with accuracy 1 ml and the time needed to pump this volume was measured by stopwatch. The division value of stopwatch equals 1 s the flow rate was defined as division of pumped value to the time required to pump it.

The comparison of calculation results and experimental data shows that the biggest difference between experimental data and calculation results occurs when the pressure drop $\Delta p = 0$. In this case the expression (16) results the half of the volume displaced by the first plate. Thus it may be concluded that difference between experimental data and calculation results takes place due to the velocity change from it maximal value to the minimum value was not considered, therefore the calculated flow rate is less than the experimental value.

5 CONCLUSION

An analytical model for linear peristaltic pump flow rate calculation in the case of laminar flow was developed. The dependence of flow rate from pressure drop was achieved. The comparison of calculation results with the experimental data shows that the difference in the flow rate values decrease with the pressure drop.

References:
[1]   Mikheev, A. Yu. *The research of characteristics and reliability improvement of peristaltic pumps*, PhD dissertation, Ufa. 2004. (in Russian)
[2]   Faraji, A.; Razavi, M. & Fatouraee, N. (2014). *Linear peristaltic pump device design*, Applied Mechanics and Materials, vol. 440, pp. 199–203.
[3]   Mermone, A.; Mazumdar, J. & Lucas, S. (2002). *A Mathematical Study of Peristaltic Transport of a Casson Fluid*, Mathematical and Computer Modelling, No. 35, pp. 895–912.
[4]   Walker, S. & Shelley, M. (2010). *Shape Optimization of Peristaltic Pumping*, Journal of Computational Physics, Feb 2010, Vol. 229, No. 4, pp. 1260–1291.
[5]   Loitsyanskiy, L. G. *Mechanics of Liquids and Gases: The textbook for higher education institutions* – 7th ed., corrected. Moscow, Drofa, 2003, p. 840.

THE INFUENCE OF TOLERANCES OF THE MAIN SIZES OF THE MULTI-STAGE CENTRIFUGAL PUMP ON THE DISPERSION OF PUMP HEAD AND PUMP EFFICIENCY

Sheypak, A. A.; Messineva, N. & Chivileva, M.

Abstract: A method of the accounting for both one-sided and double-sided tolerances for the main indicators of the multi-stage centrifugal pump is presented. Qualitative and quantitative dependences of influence of the tolerances on outer diameter for the output pump parameters are established. "Chi-square" distribution for one stage was transformed in normal distribution at considerable amounts of stages.

Keywords: *multi-stage centrifugal pump, tolerances, head, efficiency.*

1 INTRODUCTON

With mass production of centrifugal pumps due to the deviations of the sizes of the flow part being caused by manufacturing techniques, the output parameters of the product (head, efficiency) have deviations from the given values. In some cases, technology testing and refinement of details in order to supply small scatter of the machine parameters are provided. In all cases it is appropriate to have an opportunity to numerically evaluate the effect of tolerances on the manufacturing of the flow part. In the case of making the flow part using casting or electrical erosion, the tolerance usually assigned to be symmetrical. When we applied the turning or cutting treatment it is rationally assigned a tolerance to the "metal" to be one-sided. The estimation of the impact of tolerances for turbine by calculation method performed good results and allowed to move from continuous monitoring machine performance through techno-logical tests to the selected tests, significantly reducing the cost of production [1].

2 STATEMENT OF THE PROBLEM

In the simulation of technological process variations of geometrical parameters it was suggested that the geometrical dimensions have the normal distribution law [4]. Random ("pseudo-random") numbers have normal distribution law with parameters: $M = 0$, $\sigma = 1$, where M is the mathematical expectation and σ – root-mean-square (standard) deviation.

In the case of one-size tolerance it is necessary to use the distribution "Chi-square" with two degrees of freedom, that is having the following parameters $M = 2$, $\sigma = 4$.

3 THE MAIN SIZES OF FLOW PART OF A CENTRIFUGAL PUMP

Previously the centrifugal pump of console type 2K-9 [2] where the influence of the one-stage tolerances as well symmetrical ones was considered. Then the work on the basis of the multistage centrifugal pump 222ETsNAKI5A-25 (TV) was carried out. The chosen pump includes 111 wheels concluded in one section, which length is 3 meters.

The stage (Fig. 1) consists of 2 main components (the case – 1, the working wheel – 2).

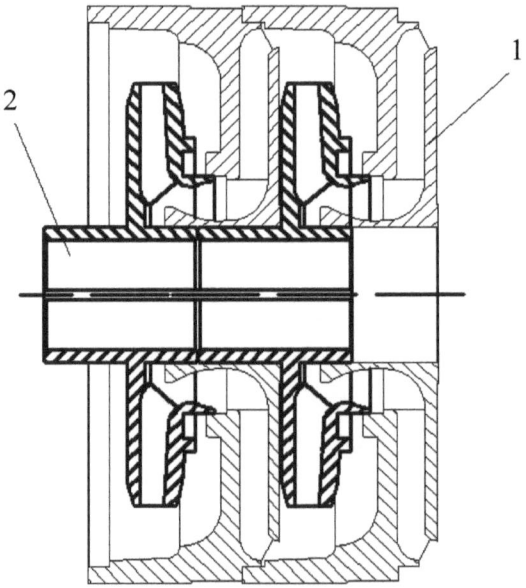

Fig. 1. Multi-stage centrifugal pump

4 INFLUENCE OF TOLERANCES OF THE MAIN SIZES OF FLOW PART OF A CENTRIFUGAL PUMP ON DISPERSION OF HEAD AND EFFICIENCY

In multi-stage pumps liquid passes consistently through some wheels mounted in one case. The multi-stage design of the pump allows to increase a pump head in comparison with one-wheeled one in so many times, how many wheels the pump has. Each working wheel with its directing device and the overflow channel forms section of the multi-stage pump. Pumped-over liquid consistently passes from one wheel into another and exits in a delivery sleeve pipe with the head proportional to number of stages or wheels.

In [3] the following expressions for head and efficiency definition are obtained. The size of a theoretical head at acceptance of a hypothesis of active radius is:

$$H_t = H_{t\infty} - \frac{u_2^2}{g}(1 - Y),$$ (1)

where $Y = 1 - 0,005\beta_2$ is the coefficient of active radius for angles ($5° < \beta_2 < 40°$), β_2 is taken in degrees along the middle line of the profile, $H_{t\infty} = v_{2u\infty}u_2/g$ is the theoretical head at infinitive amount of impeller. The total efficiency is determined by the expression:

$$\eta = \eta_o \cdot \eta_\Gamma \cdot \eta_M,$$ (2)

where η_o is the volume efficiency, η_Γ is the hydraulic efficiency, η_M is the mechanical efficiency.

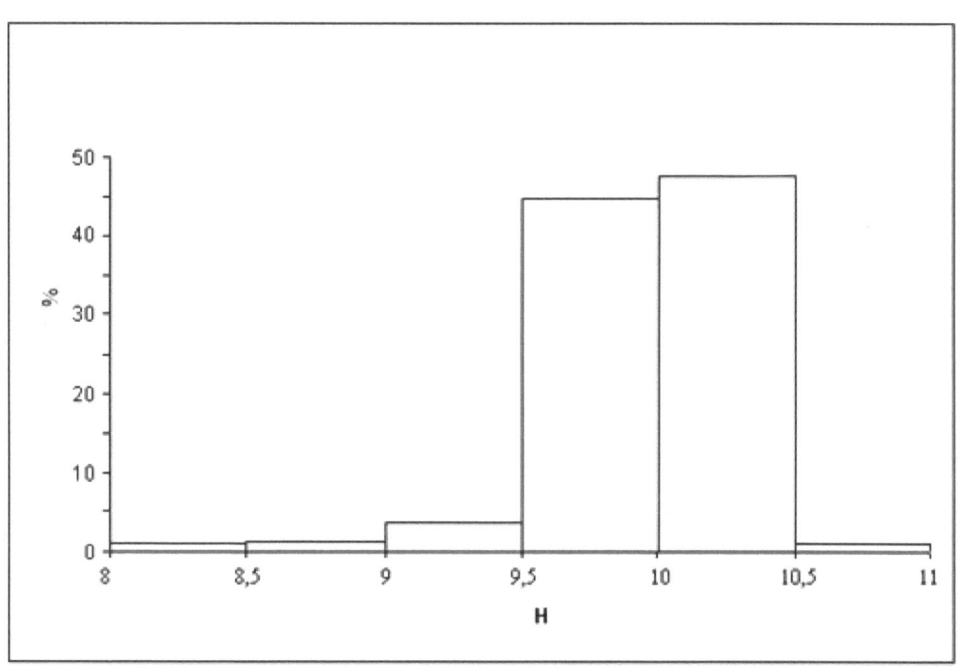

Fig. 2. Histogram of head distributions; selection is made for one wheel from 1000 values

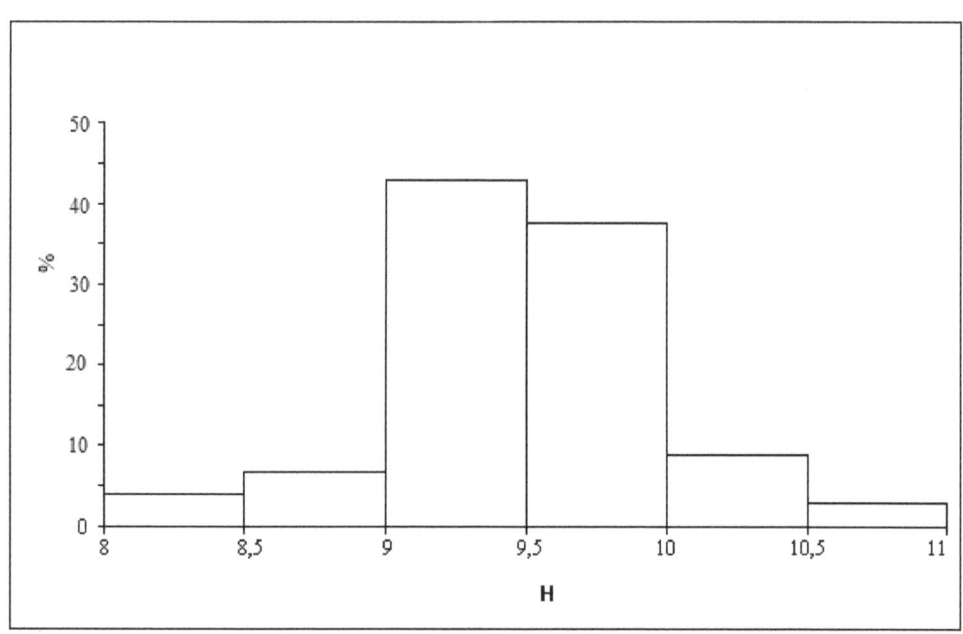

Fig. 3. Histogram of head distributions; selection is made for 111 wheels

For a more exact calculation, the analysis was carried out utilizing the obtained results on coefficient of asymmetry and excess. The results were summarized in Tables 1 and 2.

Parameters	For 1 wheel	For 111 wheels
Average	9,96	8,93

Root-mean square deviation	0,449	0,302
Excess	1,467	2,938
Selective asymmetry	0,502	0,013

Tab. 1. Verification of calculated data for head

Parameters	For 1 wheel	For 111 wheels
Average	0,662	0,66
Root-mean square deviation	0,014	0,0081
Excess	2,981	2,944
Selective asymmetry	0,167	0,11

Tab. 2. Verification of calculated data for efficiency

On the basis of the obtained results it is possible to make the conclusion about the significant difference of selective asymmetry for two options of calculation. This results from the fact that the tolerance on outer diameter is not symmetric.

5 CONCLUSION

The mathematical model for calculating the impact of tolerances for the output parameters of the multi-staged centrifugal pump is developed. Qualitative and quantitative dependences of influence of the tolerances on outer diameter for the output pump parameters are established. "Chi-square" distribution for one stage was transformed in normal distribution at considerable amounts of stages.

References:
[1] Sheypak, A. A. & Sheypak, I. et al. *Stochastic model of the centrifugal pump*, Proceedings of the 11th International Scientific Conference on Advanced Engineering, Computer Aided Design and Manufacturing, September 16 -20, 2013, Petrčane, Croatia, pp. 63-66.
[2] Sheypak, A. A. (2009). *The effects of technological sizes deviations of flow part of the pneumodynamic drive on the efficiency changes*. Mechanical Engineering and Engineering Education №4 (21), pp. 2-10.
[3] Lepeshkin, A. V.; Mikhailin, A. A. & Sheypak, A. A. *Hydraulics and hydropneumodynamic drive. Part 2. Hydraulic machinery and hydropneumodynamic drive*. Ed. by Sheypak A. A. – MSIU, 2008, p. 352.
[4] Berkov, N. A.; Minostsev, V. D. & Shishanin, O. E. Course of Higher Mathematics, Part 3 – MGIU, 2007, p. 192.

SPECIALIZED ALTERNATIVE ALGORITHM FOR DETERMINING THE LOCATION OF HYDRAULIC LIFT

Sheypak, A. A. & Novikov, P. V.

Abstract: The article describes different navigation systems, basic principles of their working, advantages and disadvantages of these systems, describes a complex system which consist of INS, Glonass and odometer. We also consider different error correction methods.

Keywords: acceleration, satellite, error model, gyro drift, odometer.

1 INTRODUCTION

The fast development of technology for the last ten years has shown a great opportunity for successful solution of different navigation tasks, with the help of different devices that can be sat on the board of any craft or outside it. These tasks are actual not only in aviation, missiles, fleet, but also on the ground transport. But the high price is a reason why people cannot use them everywhere.

2 STATEMENT OF THE PROBLEM

There are two different ways to calculate the way, inertial navigation systems (INS) and Glonass. The main idea of the inertial navigation is based on the acceleration integrations. The first integration of the vehicle acceleration provides velocity. The second integration gives vehicle position increments. The principle of Glonass working differs from INS working. Glonass is a network of about 24 satellites orbiting the Earth. Wherever you are on the planet, at least 4 satellites are visible at any time. Each one transmits information about its position and the current time at regular intervals. These signals, travelling at the velocity of light, are intercepted by the Glonass receiver, which calculates how far each satellite is located and how long it takes the messages to arrive. But this calculation has an error, caused by the clock drift. Both methods have advantages and disadvantages. INS systems collect errors during the work, and it is their main disadvantage. These errors are caused by gyro drift. The advantage of inertial systems is there autonomous working, all calculations are made on the board of the craft.

A Glonass receiver calculates its position by precisely timing the signals sent by Glonass satellites high above the Earth. Each satellite continually transmits messages that include time the message was transmitted and satellite position at time of message transmission. The receiver uses the messages to determine the time transmission of each message and calculates the distance up to each satellite using the velocity of light. Each of these distances and satellites locations defines a sphere. These distances and satellites locations are used to calculate the location of the receiver using the navigation equations. Mistakes do not accumulate during their work. The disadvantage of Glonass connects with disappearing of the signal. The universal system, which can work with all kinds of objects, does not still create. That is why the creation of system, which allows to use extra information about motion is relevant. The source of this information can be odometer – the device which counts the way. Odometer counts the rotation of wheel and conversion this to the value of the traversed path. There are several automobile systems, which include odometer. Unfortunately most of these systems exist like a prototype. High price and low accuracy do not allow serial production. That is why the creation of a low cost system able to determine the position of the object with a given accuracy is still topical. INS errors determined by the accuracy of using sensors. The gyro drift is a very important parameter. If we use these systems on Earth we can get information about velocity using odometer. Thus there is no need to count thise velocity using INS systems.

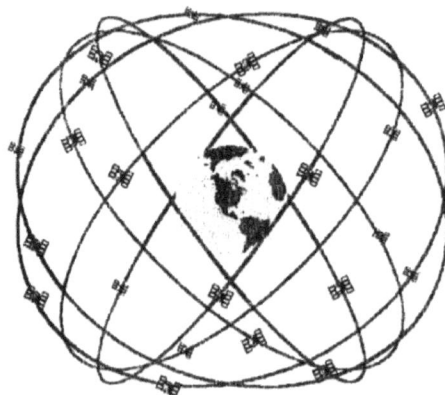

Fig. 1. Configuration of orbital grouping system Glonass

Engineers attempted to compare two systems and make them work alternately. But the high price of the device does not allow to produce them. The main goal now is to create a complex cheap system that consists of INS and Glonass with less accurate sensors and creation of data processing algorithm providing the specified accuracy. The main idea of the working system is alternately switching of INS and Glonass. Error correction of INS by uses classic algorithm from Glonass receiver can be done with the help of the closed method. It is based on the application of additional control actions to the real or imaginary system platform. This system can be done only if for arbitrary motion can be allocated in the output values of the inertial velocity δV_e and δV_n. To avoid errors, we can apply external sources of information about the speed. If we have eastern and northern components of the velocity of the object due to Glonass, then up to its own satellite system errors:

$$\delta V_e = V_e - V_E^{Gl}, \quad \delta V_n = V_n - V_n^{Gl}. \tag{1}$$

Damping-$K_1\delta V$ control signal is input to the velocity calculation unit of the navigation algorithm. The single channel error model has the form (E – channel):

$$\delta \dot{V}_e = -g\Phi_N + \delta f_E - K_1 V_E, \quad \dot{\Phi}_N = \frac{\partial V_e}{R} + \delta \omega_N + K_2 \delta V_E. \tag{2}$$

Differentiating the first equation and substituting the second in his right part, we obtain the variation of error δV_n:

$$\delta \ddot{V}_{\mathrm{E}} + K_1 \delta \dot{V}_{\mathrm{E}} + \left(\frac{g}{R} + K_2 g \right) \delta V_{\mathrm{E}} = -g \delta \omega_{\mathrm{N}} + \delta f_{\mathrm{E}} . \tag{3}$$

From the last equation it follows that the error in the output data decreases subsequently time, and the frequency is

$$\omega_0 = \sqrt{v^2 + K_2 g} \ \square\square \ v. \tag{4}$$

The closed method uses special algorithms for estimating errors of uncorrectable (autonomous) inertial system which can be based on a priori information about their character, as well as the measurement of object motion parameters using external devices, including Glonass receiver and odometer. The most common and effective algorithm of this kind is the optimal Kalman filter. Let us consider any linear system that can be described by certain values, usually inaccessible for the measurement. These values are elements of state vector $\overline{x}_k$ adopted a linear variation of the system (model):

$$\overline{x}_{k+1} = \Phi \overline{x}_k + G \overline{w}_k , \tag{5}$$

where Φ – transition matrix, G – input matrix, $\overline{w}_k = \overline{w}(t_k)$ – input white noise with covariance matrix Q.

The part of the state vector components or their linear combinations are directly observable according to the equation:

$$\overline{z}_k = H \overline{x}_k + \overline{v}_k , \tag{6}$$

where $\overline{z}_k$ – measurement vector; $\overline{v}_k$ – measurement white noise with zero meaning and known covariance matrix R.

Kalman logic allows obtaining the best estimation ξ_k of the system, based on the measurement with errors. The Kalman filter provides minimization of the mean-square error.

$$J_k = \mathrm{TrM} \left[\left(\overline{x}_k - \xi_k \right) \left(\overline{x}_k - \xi_k \right)^{\mathrm{T}} \right] \to \min . \tag{7}$$

At the first stage of applying the algorithm creates a priori estimate $\xi_{k+1|k}$ of the state vector of the system based on the assumption that the adopted model is accurate and contains no input noise:

$$\xi_{k+1|k} = \Phi \xi_k \tag{8}$$

When the next value of the measurement vector $\overline{z}_{k+1}$ is obtained, it is possible to specify an a priori estimate by introducing feedback changes with matrix K_{k+1}:

$$\xi_{k+1} = \xi_{k+1|k} + K_{k+1} \left(\overline{z}_{k+1} - H \xi_{k+1|k} \right) . \tag{9}$$

The second stage of the application filter algorithm includes finding the optimal value of the matrix K_{k+1} in accordance with the selected criteria:

$$\frac{dJ_{k+1}}{dK_{k+1}} = 0 , \quad K_{k+1} = P_{k+1|k} H^{\mathrm{T}} \left(H P_{k+1|k} H^{\mathrm{T}} + R \right)^{-1} , \tag{10}$$

where $P_{k+1|k} = M \left[\left(\overline{x}_{k+1} - \xi_{k|k+1} \right) \left(\overline{x}_{k+1} - \xi_{k|k+1} \right)^{\mathrm{T}} \right]$ is the covariance matrix of estimation errors, which can be calculated as:

$$P_{k+1|k} = \Phi P_k \Phi^{\mathrm{T}} + G Q G^{\mathrm{T}} . \tag{11}$$

Then we can calculate the covariance matrix:

$$P_{k+1} = (I - K_{k+1}) P_{k+1|k} \tag{12}$$

If we cannot find the measurement vector $\overline{z}_{k+1}$ then Kalman filter is put on prediction mode. For example, take $R = \infty$ and then, according to the above described sequence of operations, the state vector estimation is performed using only a single model, and coincides with the a posteriori estimation of the a priori:

$$\xi_{k+1} = \Phi \xi_k . \tag{13}$$

Concurrently with the main estimation function Kalman filter allows to smooth the noise contained in the measurement vector $\overline{z}_{k+1}$. The principle of open inertial correction relies on the definition of its errors. So, for the eastern channel platform inertial obtain:

$$\begin{bmatrix} \delta E \\ \delta V_E \\ \Phi_N \\ \delta \omega_N \end{bmatrix}_{k+1} = \begin{bmatrix} 1 & T & 0 & 0 \\ 0 & 1 & -gT & 0 \\ 0 & T/R & 1 & T \\ 0 & 0 & 0 & 1 \end{bmatrix} \begin{bmatrix} \delta E \\ \delta V_E \\ \Phi_N \\ \delta \omega_N \end{bmatrix}_k + \begin{bmatrix} 0 \\ 0 \\ 0 \\ T \end{bmatrix} w_k , \tag{14}$$

$$\begin{bmatrix} E & -E^{Gl} \\ V_E & -V_E^{Gl} \end{bmatrix}_{k+1} = \begin{bmatrix} 1 & 0 & 0 & 0 \\ 0 & 1 & 0 & 0 \end{bmatrix} \begin{bmatrix} \delta E \\ \delta V_E \\ \Phi_N \\ \delta \omega_N \end{bmatrix}_{k+1} + \begin{bmatrix} \delta E^{Gl} \\ \delta V_E^{Gl} \end{bmatrix}. \tag{15}$$

Knowing at each step of the calculation the value E^{Gl}, V_E^{Gl} and applying the Kalman filter, we can estimate the errors of the inertial system, including those that are not available for direct measurement (in particular, the deviation Φ_N platform from the horizontal plane and angular velocity $\delta \omega_N$ its drift). These estimates are used to amend the output readings of the coordinates $\overline{R} = [E, N, U_p]^T$, velocity $\overline{V} = [V_E, V_N, V_{Up}]^T$ and the orientation angles $\overline{\Psi} = [\gamma, \vartheta, H]^T$.

3 CONCLUSION

This paper proposes an alternative approach procedure to evaluate location settings land mobile objects that require precise positioning. The proposed system fundamentally differs from the existing one because it is based on sensitive elements having a small size and a low market price. The proposed algorithm analysis of the system was tested in actual operating conditions.

References:
[1] Salychev, O. S. *Applied Inertial Navigation: Problems and Solutions* – M.: BMSTU Press, 2004.
[2] Vlasenko, A. *Integrated gyroscopes iMEMS – angular velocity sensors*, Analog Devices firm / Electronic Components, 2003.
[3] Hemerly, E. M. & Schad, V. R. *Implementation of a GPS/INS/Odometer*, 2008.

GREEN FUNCTIONAL SUBASSEMBLIES FOR CONCEPTUAL DESIGN

Takács, Á.

Abstract: The paper deals with the elements of environmentally friendly design. It analyzes how the most popular techniques are operating and gives a suggestion for implementing introduced techniques. According to the analysis it proposes to make a list of green functional subassemblies that is suitable for adapting to a software that helps designers during conceptual design process.

Keywords: design theory, methodology, Design for Environment.

1 INTRODUCTION

It is not easy to shortly define environmentally friendly design. According to the several components it has, it is quite a complex process. According to Zilahy [1] environmentally friendly design systematically concentrates to the environmentally impacts that are potentially coming in the fore during the whole life-cycle of products and services, and to reduce or eliminate these expectable impacts still in the design process.

Orbán [2] defines DFE as a design that minimalizes the undesired impacts for the nature (DFE = design for environment). DFE is the necessity of the developed product causes the less harmful impact on the environment that is an ever-growing claim of today.

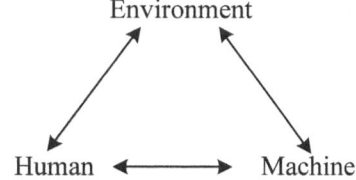

Fig. 1. Machine-human-environment cycle

Due to the literature of the field environmentally friendly design or DFE or Green design or eco design mean only the protection of the nature do not pay any attention to the protection of the human that is a component of the green environment, only indirectly referring to it. It is essential to notice that the man only as a designer but also as the part of the green environment appears in the machine-human-environment cycle. The elements of the cycle are interrelationship continuously, as Figure 1 shows. So the human designs for itself and for the environment as well. Machine has the effect for the human and for the environment too. The environment also has the impact for the human and for the machine. So environment means not only the nature over the office, the factory, but the direct environment of the human where it works, so the workplace. As for the further researches it would be practical to mention and analyse ergonomics as the element of the environmentally friendly design.

2 TOOLS OF DFE

DFE elements:

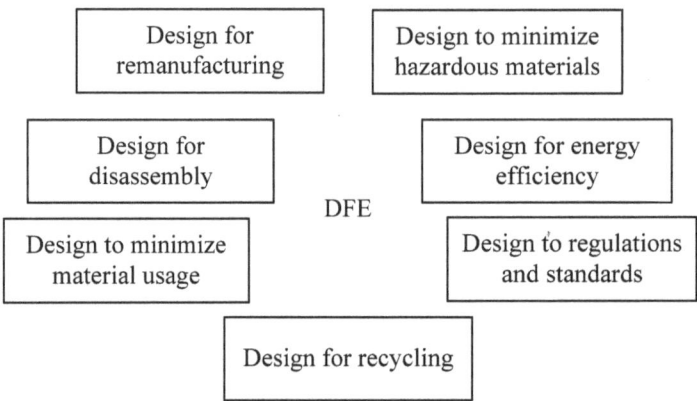

Fig. 2. Design for the Environment [3]

Dfx, or design according to a given viewpoint can be any formal period of the design process, or any important aspect that can be followed during the whole design activity as the main principle. Dfx is an enormous set of design principles that is really hard to describe, because this set is increasing day-by-day. Scientists define more and more principles, and for those principles methods are also created. These methods denote or can denote the adaption of Dfx techniques to computer. DFE, that is Design for the Environment is collecting the aspects of environmentally friendly design. As it is shown in Figure 2 it consists of seven essential areas. According to different aspects these can be divided into other different principles. This figure also confirms why it is so complicated to collect all the Dfx techniques and to group them.

3 ERGONOMICS

MacLeod [4,5] defined twelve principles that can help the designer's work during the design process to create a machine, tool, equipment or product that ensures comforttable work for the user. These principles are general, but give significant help during design. David Ridyard determined five main territories within he declared several design principles for neutral postures. The aim is to ensure these normal positions.

4 CONCEPTUAL DESIGN

Figure 3 [6] focuses on the main scope of researches; it summarises the phase of the conceptual design. Suggested method introduced by Figure 3 implies a relatively simple algorithm, so the process is adaptable for computer. The introduced method consists of a quantity and a quality line. Quantity line makes it possible the designer could pay attention on more aspects, so more functional subassemblies and this way more solutions. Quality line valuates solutions according to different view-points and tightens the solution-space, optionally for one proper solution. In the modern World of our days it significantly facilitates the task of the engineering designer.

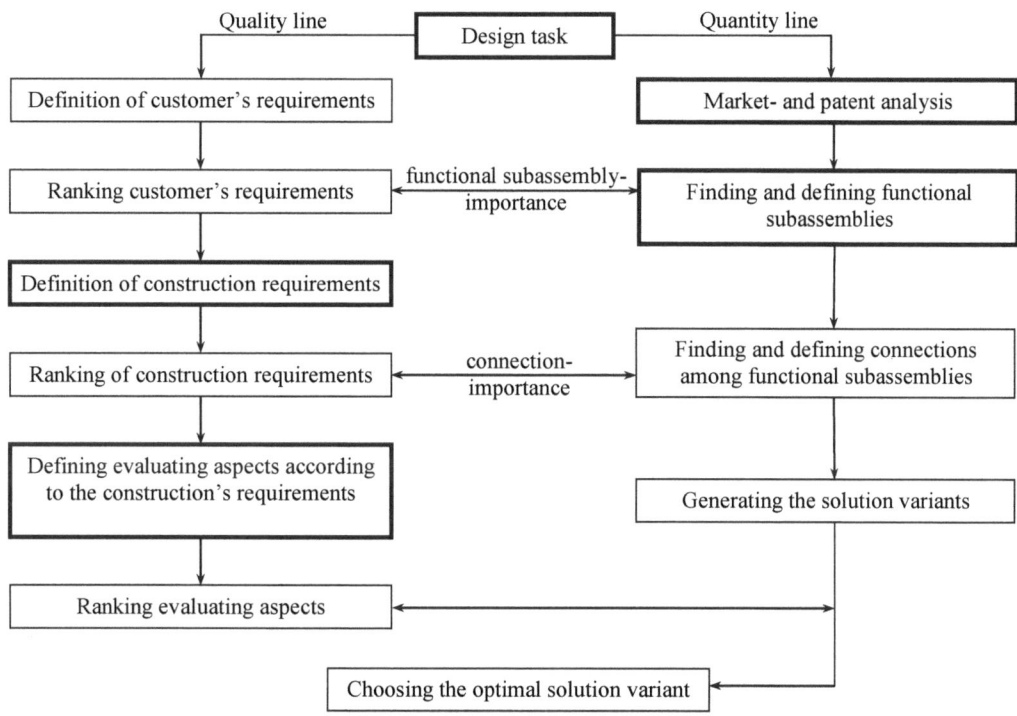

Fig. 3. Conceptual design process – a suggestion

As it is shown in Figure 3, certain steps of the conceptual design process are indicated by thick line. These are the steps where the creativity of the designer appears and the designer has to pay attention to the given circumstances, rules, laws and in these steps the tools of the DFE can also be taken into consideration. Right now, on this stage of the research it cannot be defined by numbers how effective it is. Quantification would be easier if a catalogue have been composed that would collect those functional subassemblies that have some kind of 'green' effect, and take one of the DFE tool as a basis.

Green functional subassemblies have to be determined in a separate group to take into account the expectable effects on the environment with the help of an existing software operating on the basis of the process introduced in Figure 3. For example solar battery different types of filters, led lightning can be green functional subassembly. These should be ranked on the basis of their expectable effect on the environment, but designers' claims also have to be kept in front of the eye. According to the importance order of green functional subassemblies a diagram can be composed that ranks the promising solutions due to the number of green functional subassemblies can be found in them. So the diagram in Figure 4 arises.

The diagram shows, which preliminary determined and ranked green functional subassemblies are chosen to compose the different solutions. So solution i (S_i) contains all the green functional subassemblies signed in diagram (Fy, Fh, Fa, Fv, Fj) among them the strongest one that has the highest environmental effect is functional subassembly j, while the weakest one is functional subassembly y that has the less effect on the environment. This weakness means that the given functional subassembly assists the less the environmentally friendly design. So solution S_i is the most environmentally friendly according to Figure 4.

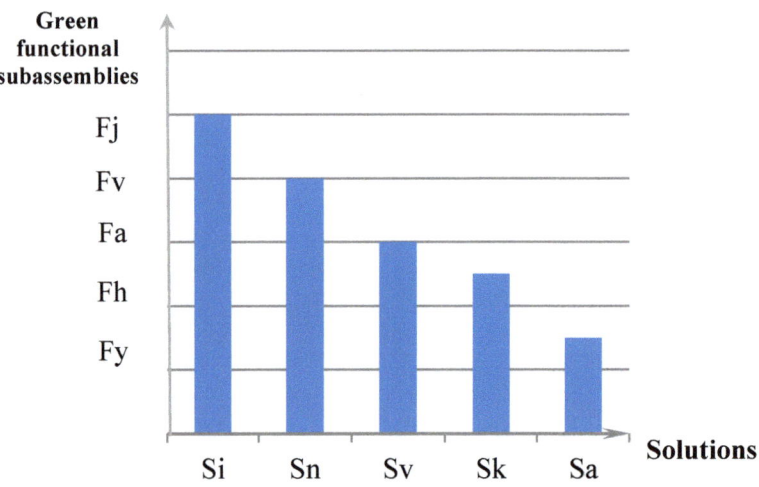

Fig. 4. Ranking solutions on the basis of green functional subassemblies

5 SUMMARY

On the basis of the above introduced principles defining a tip-list is in process that takes not only environmentally friendly design but ergonomics as well into consideration. And on this basis gives suggestions to the designer during the conceptual design phase while making the list of functional subassemblies by suggesting ones from the built-in ones. Further task is to develop the subassembly-kit and the tip-list.

Acknowledgement:
This research was supported by the European Union and the State of Hungary, co-financed by the European Social Fund in the framework of TÁMOP-4.2.4.A/2-11/1-2012-0001 'National Excellence Program'.

References:
[1] Zilahy, Gy. *Tisztább termelés, Budapesti Corvinus Egyetem*, HEFOP-3.3.1., előadásfóliák.
[2] Orbán, F. *Környezetszempontú tervezés*, Budapesti Műszaki és Gazdasági Egyetem, HEFOP-3.3.1., előadásfóliák.
[3] Otto, K. & Wood, K. *Product Design – Techniques in Reverse Engineering and New Product Development*, ISBN 9780130212719, Prentice Hall, 2008.
[4] MacLeod, D. *The Ergonomics Kit for General Industry*, ISBN 1280546115, ebook, CRC Press, 2006.
[5] MacLeod, D. *The Rules of Work – A Practical Engineering Guide to Ergonomics*, ISBN 1560328851, ebook, CRC Press, 2000.
[6] Takács, Á: *Számítógéppel segített koncepcionális tervezési módszer*, doktori (PhD.) disszertáció, Miskolc, 2010.

AN ANALYSIS OF THE LOADING CAPACITY OF INTERNAL HIGH CONTACT RATIO GEARS

Vrcan, Ž. & Lovrin, N.

Abstract: An increase in gearbox power ratings has been followed by a reduction in gearbox size and weight. This is possible because of high contact ratio (HCR) gears, because these gears distribute the load between at least two tooth pairs in mesh. Therefore the gears can be made smaller and lighter. The loading capacity of internal HCR gears has not been thoroughly researched yet. The method outlined in ISO 6336 is analyzed in this paper, and alternative calculation methods have been proposed, and the results compared.

Keywords: *internal gears, high contact ratio, loading capacity, tooth root stress, tooth flank stress.*

1 INTRODUCTION

High contact ratio (HCR) gears are non-standard involute spur gears with a transverse contact ratio higher than two ($\varepsilon_\alpha > 2$). The tooth pairs of HCR gears are always in simultaneous double or triple contact, unlike standard involute gears, where the tooth pairs are in single or simultaneous double contact.

Therefore, HCR gears will have increased load capacity, and reduced gear size in relation to a standard involute gear carrying the same load.

Theoretically speaking, an absolutely rigid, perfectly machined conventional involute gear will be alternatively loaded along the path of contact by 50% and 100% of the total normal force F_{bt}, while a HCR gear would be respectively loaded by 33% and 50%. However, due to elastic deformation of actual gear teeth, the maximal normal force F_{bti} on a tooth along the path of contact A-G will exceed 50% of the total normal force F_{bt}. [1,2].

2 LOAD CAPACITY CALCULATION

Tooth root stress calculation by ISO 6336 [3] assumes that the load is acting on the tooth in the outer single point contact, while for HCR gears it is assumed to act in the internal triple point contact. Analysis of the gear mesh in [1] has shown that the internal triple point contact for HCR wheel gears in the case of tooth root calculation is at point E. Tooth root stress is calculated using the equation (1):

$$\sigma_{F0} = \frac{F_t}{b \cdot m_n} Y_F Y_S Y_\beta Y_B Y_{DT},\tag{1}$$

where Y_F is the tooth form factor, Y_S is the stress correction factor, Y_β is the helix angle factor, Y_B is the gear rim thickness factor, and Y_{DT} is the deep tooth factor.

Examination of (1) has shown that ISO 6336 attempts to model the actual meshing conditions of internal HCR gears by reducing root stress via the deep tooth factor Y_{DT}.

Tooth flank stress calculation by ISO 6336 [3] states that the stress will be calculated for the pinion internal double contact point E without further elaboration. Equation (2) is used to calculate tooth flank stress (2):

$$\sigma_{H0} = Z_H Z_E Z_\varepsilon Z_\beta \sqrt{\frac{F_t}{d_1 b} \frac{u+1}{u}} , \qquad (2)$$

where Z_H is the zone factor, Z_E is the elasticity factor, Z_β is the helix angle factor, Z_ε is the contact ratio factor, and u is the gear ratio. Analysis of (2) has shown that the actual meshing conditions of HCR gear flanks are partially modelled by means of the zone factor Z_H, and the contact ratio factor Z_ε. The gear ratio factor u also contributes to the representation of convex to concave tooth flank contact characteristic to internal gears. Additionally, (2) is restricted to transverse contact ratios $2 < \varepsilon_\alpha \leq 2,5$, meaning that a more accurate calculation method is needed.

Based on the work of [4], a new method for the calculation of tooth root stress deflection was proposed in [5, 6, 7]. This method is based on finite element research, and places the point of maximum tooth root stress at the point in which the root fillet tangent closes an angle of 45° to the tooth centerline.

This point is also used to establish the reference tooth thickness s and loading arm length l for tooth root stress calculations, unlike the ISO 6336 method which uses the point in which the root fillet tangent closes an angle of 60° to the tooth centerline. The tangent angle of 45° to the tooth centerline was selected as tooth root cracking of internal gears almost always occurs at this point of the tooth root curve [5].

This method of stress calculation is based on the analysis of the tooth shape by conformal mapping, and takes into account the stresses occuring in the tooth root, notably the nominal bending stress σ_{Nb1}, pure moment bending stress σ_{Nb2}, point loading compressive stress σ_{Nc}, and point loading shear stress τ_N.

The internal gear tooth root stress may then be calculated by means of equation (3):

$$\sigma_F = \left(1 + 0,046 \frac{s}{\rho}\right) \cdot \left(0,67\sigma_{Nb} + 0,48\sqrt{\sigma_{Nb}^2 + 36\tau_N^2} + 1,14\sigma_{Nc}\right), \qquad (3)$$

where σ_{Nb} is the combined bending stress, and ρ is the radius of curvature at the point in which the root fillet tangent closes an angle of 45° to the tooth centerline.

An analysis of equation (2) has shown that the curvature of the tooth flanks in contact is not taken into consideration. Therefore, equation (2) was recalculated into equation (4):

$$\sigma_{H0} = Z_\varepsilon \sqrt{\frac{F_{bt} E}{2\pi\rho_Y b\left(1-v^2\right)}} , \qquad (4)$$

where ρ_Y is the radius of curvature at the point of tooth flank contact, and v is Poisson's ratio. Equation (4) is based on Hertzian contact stress theory, and is therefore suitable for contact stress calculation for internal gears.

3 COMPARISON OF RESULTS

The proposed calculation methods have been evaluated on a gear pair with the following characteristics: pressure angle $\alpha = 20°$, module $m = 22$ mm, center distance $a = -726$ mm, teeth numbers $z_{1,2} = 22 / -88$, tooth dedendum factor $h_{a01,2}* = 1,5$ m, tip radius factor $\rho_{a01,2}{}^* = 0{,}341/0{,}259$, addendum modification coefficients $x_{1,2} = 0$, reference diameter $d_{1,2} = 484 \, / \, {-}1936$ mm, tip diameter $d_{a1,2} = 545{,}6/{-}1886{,}3$ mm, base diameter $d_{b1,2} = 454{,}81/{-}1819{,}25$ mm, contact ratio $\varepsilon_{\alpha} = 2{,}306$. The total tangential force acting on the gear teeth in mesh was $F_{bt} = 26500$ N. Finite element analysis was used to determine the load distribution during the mesh. The results are shown in Figures 1 and 2.

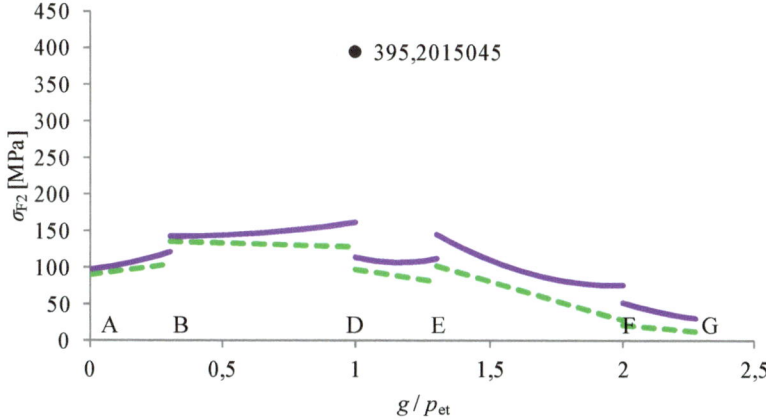

Fig. 1. IHCR gear tooth root stress σ_{F2} during the mesh

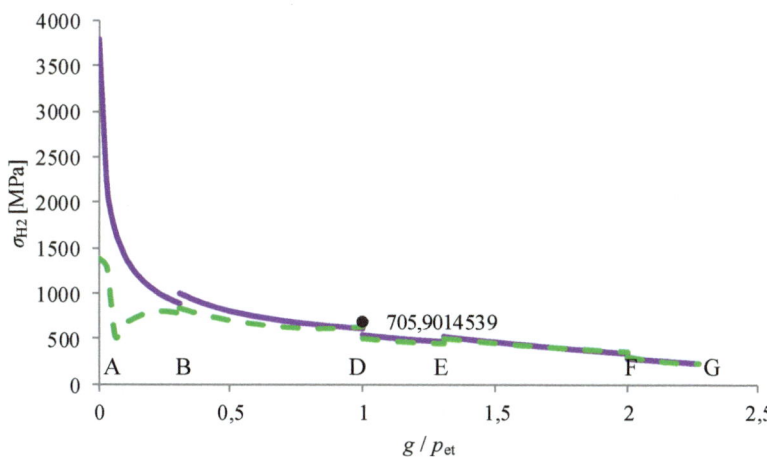

Fig. 2. IHCR tooth flank stress σ_{H2} during the mesh

The tooth root stress during the mesh of the tested IHCR gear is shown in Figure 1. Tooth root stress calculated using FEM (solid line) increases from the beginning of the first triple contact at meshing point A towards the beginning of the first double contact at point B. Tooth root stress then reaches its apex value at the end of the first double contact, and then decreases at the beginning of the second triple contact at meshing point D. Tooth root stress increases again at the beginning of the second double contact

69

period at meshing point E, before finally decreasing towards the third triple contact period and meshing points F and G.

Tooth root stress according to ISO 6336 was calculated for meshing point D (dot), while root stress values according to (3) (dashed line) were calculated for the whole path of contact using finite element analysis derived load distribution data.

The tooth flank stress during the mesh of the tested IHCR gear is shown in Figure 2. Tooth flank stress calculated using FEM (dashed line) decreases from its apex at the beginning of the first triple contact at meshing point A towards a local minimum during the first triple contact period. Tooth flank stress then rises towards the beginning of the first double contact at point B, before gradually decreasing during the rest of the contact. Tooth flank stress according to ISO 6336 was calculated for meshing point D (dot), while flank stress values according to (4) (solid line) were calculated for the whole path of contact using finite element analysis derived load distribution data.

4 CONCLUSION

Analysis of the tooth root stress data displayed in Figure 1 has shown that the values calculated using (3) follow the results obtained by finite element analysis very well. Tooth root stress according to ISO 6336 is much higher than both of those values.

Analysis of tooth flank stress data displayed in Figure 2 has shown that the values calculated using (4) follow the results obtained by finite element analysis very well beyond meshing point B. This difference can be explained by the influence of radii of curvature of gear tooth flanks at the point of contact on the results of (4). Tooth flank stress according to ISO 6336 is higher than both calculated values for meshing point D.

Therefore, it can be summed up that that gears calculated according to ISO 6336 will have excessively high safety factors, and that calculations according to (3) and (4) will result in realistically dimensioned gears which will still fulfil all safety factor requirements.

References:
[1] Lovrin, N. (2001). *Load Capacity Analysis of the High Transverse Contact Ratio Involute Gearing*, (in Croatian), Thesis. Faculty of Engineering, Rijeka, Croatia.
[2] Vrcan, Ž. (2014). *A Contribution to the Research of the Loading Capacity of Internal Involute High Contact Ratio Gears*, (in Croatian), Thesis. Faculty of Engineering, Rijeka, Croatia.
[3] ISO 6336 (2006), *Calculation of load capacity of spur and helical gears,* ISO, Geneva.
[4] Terauchi, Y. & Nagamura, K. (1981). *Study on Deflection of Spur Gear Teeth*, Bulletin of JSME, Vol. 24, No. 188, pp. 447-452.
[5] Oda, S. & Miyachika, K. (1986). *Root Stress Analysis of Internal Spur Gears by Theory of Elasticity*, Bulletin of JSME, Vol. 29, No. 250, pp. 1287-1293.
[6] Oda, S. & Miyachika, K. (1986). *Practical Formula for True Root Stress of Internal Spur Gear Tooth*, Bulletin of JSME, Vol. 29, No. 252, pp. 1934-1939.
[7] Oda, S. & Miyachika, K. (1991), *Bending Strength of Internal Spur Gears*, Proceedings of the MPT' 91 JSME International Conference on Motion and Power Transmissions, pp. 781-786.

Authors Index

ORGANIZING COMMITTEE ADDRESS:

Information, registration and accommodation

Congress Agency Revelin
Kolavići 5, HR-51414 Ičići, Croatia

Phone: + 385 (0) 51 299 400, 299 410
Fax: + 385 (0) 51 299 420
http://www.revelin.hr

Submission of papers

Prof. Boris Obsieger, D.Sc.
Faculty of Engineering
Vukovarska 58, HR-51000 Rijeka, Croatia

Phone: +385 (0) 51 651 531
Fax: +385 (0) 51 651 532
e-mail: cadam@edu-point.eu

Instructions for authors

http://cadam.edu-point.eu

www.ingramcontent.com/pod-product-compliance
Lightning Source LLC
Chambersburg PA
CBHW050744180526
45159CB00003B/1347